Chemical Reactions

Stoichiometry and Beyond

First Edition

By Dr. John A. Olson
Baylor University

Bassim Hamadeh, Publisher
Christopher Foster, Vice President
Michael Simpson, Vice President of Acquisitions
Jessica Knott, Managing Editor
Stephen Milano, Creative Director
Kevin Fahey, Cognella Marketing Program Manager
Rose Tawy, Acquisitions Editor
Jamie Giganti, Project Editor
Erin Escobar, Licensing Associate

Copyright © 2012 by University Readers, Inc. All rights reserved. No part of this publication may be reprinted, reproduced, transmitted, or utilized in any form or by any electronic, mechanical, or other means, now known or hereafter invented, including photocopying, microfilming, and recording, or in any information retrieval system without the written permission of University Readers, Inc.

First published in the United States of America in 2012 by University Readers, Inc.

Trademark Notice: Product or corporate names may be trademarks or registered trademarks, and are used only for identification and explanation without intent to infringe.

16 15 14 13 12   1 2 3 4 5

Printed in the United States of America

ISBN: 978-1-60927-426-9

www.cognella.com 800.200.3908

Contents

Chapter 1: The Experimental Laws, Dalton's Theory, and Relative Atomic Masses

Introduction	1
The Experimental Laws	2
Dalton's Theory	4
The Significance of Relative Atomic Masses	13

Chapter 2: The Average Atomic Mass, the Molar Mass, and Empirical and Molecular Formulas

Introduction	17
The Average Atomic Mass	22
Dalton's Theory	26
The Mole Concept	30
Empirical Formulas	37
A Prelude to Stoichiometry	47

Chapter 3: Stoichiometry

Introduction	51
Stoichiometry for Masses	53
Stoichiometry for Solutes	70
Stoichiometry for Gases	94

Chapter 4: Chemical Equilibrium

Introduction	105
Dynamic Equilibrium and the Equilibrium Expression	105
General Relationships for Equilibrium Constants	109
The Reaction Quotient	113
The Equilibrium Format	115
Calculator Evaluation of the Equilibrium Expression	136
Applications of Le Châtelier's Principle	159

Index	163

Preface

Much of the subject of Chemistry that deals with macroscopic systems consists of two general questions. The first is "What is matter?" The second is "How does matter behave?" The first chapter and the beginning of the second chapter is primarily concerned with the first question. The rest of this material is concerned with the second question. The behavior of matter can deal with two types of changes. The first are physical changes where the composition of the matter does not change. He second are chemical changes where the composition of the matter does change and these are the primary focus of much of this material.

One of the primary goals here is to take different areas of chemistry that involve chemical changes and look at them the same way. Since one must solve problems in these areas this means that all problems are looked at in the same way. Many people are taught to use a rote method for solving problems. Learn how to solve a variety of problems. When a new problem is encountered try to recognize it so that the appropriate technique can be used. The rote method begins with a very careful reading of the problem. This illustrates that one has only one question in mind, which is "Do I know how to solve this problem?" If the answer is yes all is well but if it is no this approach could completely fail. Rather than use the rote method a different method based on asking three questions that can be answered is used. This in itself is of little help but a general method of recording the answers called a format is also used. What actually occurs is that all problems are looked at the same way by asking the three questions. Then the format changes the problem that could be a paragraph long into something that is very familiar and this completely avoids any recognition. Once you learn how to move around in the format the solution of problems becomes much more straightforward.

The format is a way to link together the chemical change, the lab measurements and the world of thinking called the mole world. Much of the focus here is on the structure of the format and the general rules for moving around in it. This is often done by using a reaction with two reactants called A and B with corresponding stoichiometric coefficients a and b and with two products called C and D with corresponding stoichiometric coefficients c and d. There are numerous exercises using this reaction with well defined values of a, b, c and d. If you would prefer to have actual chemicals you can use any general chemistry textbook as a source of problems.

Finally although this not explicitly stated in this material I like to use the "Law of Laziness" to introduce new material. This law is based on doing as little as possible to achieve an objective. When this law is used an objective is accomplished without introducing any new material. Then some new material is introduced. Finally the same objective is accomplished by using the new material. This clearly justifies why the new is useful because it simplifies achieving the objective. This is used in empirical formulas, in the prelude to stoichiometry and in chemical equilibrium.

CHAPTER 1
THE EXPERIMENTAL LAWS, DALTON'S THEORY AND RELATIVE ATOMIC MASSES

Introduction

A fundamental question that has been asked for many millennia is "What is matter?". One would like to know whether there are fundamental building blocks that make up all mater and if so what they are. The problem with finding an answer from direct observation with only "primitive" laboratory equipment can be seen by considering a piece of paper. Tearing it in half gives two smaller pieces of paper. Repeating this process again and again just gives smaller pieces of paper. Even with a microscope, scalpel and tweezers one would get smaller pieces of paper until the piece became too small to observe. In other words, the answer to "Are there fundamental building blocks that make up paper?" can't be answered by direct observation. An answer to this question will therefore be in the form of a theory (intelligent guess), which is a product of the human mind.

Before 3000BC a one building block theory was recorded in Babylonia that said all matter was made up of water. Further theories were recorded in Greece between 600BC and 300BC. One of the more notable ones is the theory by Aristotle that assumed there were four building blocks (earth, air, fire and water) and four qualities (hot, cold, moist and dry). Matter including the building blocks could be changed into other matter by the addition or removal of the qualities. For example, water could be changed into air by adding hot. Another notable theory is the atomic theory developed by Leucippos and recorded by Democritus which said that the building blocks of matter were atoms and changes of matter was due to redistributions of atoms. Although there were many theories there was no way to establish with certainty which theory was correct.

The study of matter preceding modern Chemistry is known as Alchemy. Beginning is Alexandria in about 300BC it assumed that Aristotle's theory was correct. It was practiced throughout the world until the early nineteenth century in Europe. With the choice of Aristotle's theory which has only four building blocks and four qualities, anything should be able to be accomplished through trial and error in a laboratory. Their lofty objectives were to make gold or silver out of common metals such as lead, the creation of an elixir that would cure all diseases and give everlasting life and the creation of a universal solvent.

The Scientific Revolution began during the seventeenth century. Of particular notice is the introduction of the Scientific Method by Sir Francis Bacon. Indisputable facts of nature are established using planned, reproducible observations of nature called experiments. Generalizations of related experiments give laws of nature, which are also indisputable facts. A law however does not provide an explanation for the facts. Explanations are given in the form of theories that, since they are products of the human mind, can't be proved. A theory explains the corresponding laws and is capable of making predictions of experiments that have not been done. If the result of the experiment and the prediction of the theory agree, the theory gains credibility. If the result of the experiment and the prediction of the theory do not agree the theory must be modified to make the correct prediction.

The Chemical Revolution began in the late 1700's and a major contributor was Antoine Lavoiser in France. He developed the oxygen theory of combustion, introduced the modern concept of an element and listed over thirty elements, changed chemical nomenclature and wrote a text "Elements of Chemistry" that helped other chemists use his ideas. He was a strong advocate of the scientific method.

With this background we are now ready to introduce the laws preceding Dalton's Theory, which are the Law of Conservation of Mass and the Law of Constant Composition. Dalton's Theory is then introduced and it is used to explain the laws and to develop another law called the Law of Multiple Proportions. We will then consider chemical formulas of gaseous compounds and the relative masses of the elements. We will also show how relative masses can be used to do stoichiometry.

The Experimental Laws

Antoine Lavoisier studied numerous chemical reactions and accurately measured the mass both before and after the reaction took place. In his textbook "Elements of Chemistry" he explicitly stated the general results of these measurements known as the Law of Conservation of Mass. This law states that the total mass before the reaction occurs is the same as the total mass after the reaction occurs. For a reaction of the form

$$\text{Reactants} \longrightarrow \text{Products}$$
$$\text{Lab} \quad m_{React} \quad\quad\quad m_{Prod}$$

where m_{React} and m_{Prod} is the total mass of the reactants and the total mass of the products respectively, this law gives

$$m_{React} = m_{Prod} \quad . \tag{1}$$

This law can be very useful for reactions that involve gases. For example, if a metal is heated with excess oxygen, a salt like solid is formed. In the form of a reaction one has

$$\text{metal} + \text{oxygen} \longrightarrow \text{salt}$$
$$\text{Lab} \quad m_{metal} \quad\quad m_{oxy} \quad\quad\quad m_{salt}$$

and

$$m_{metal} + m_{oxy} = m_{salt} \quad . \tag{2}$$

The metal and salt are both solids and their masses can be easily determined on a scale. However, oxygen is a gas, which can't be directly weighed on a scale. With the Law of Conservation of Mass one has from Eq. (2) that

$$m_{oxy} = m_{salt} - m_{metal}$$

so that the mass of the gas does not have to be measured directly.

In 1799, Joseph Proust compared many natural compounds found in nature to the ones he prepared in his laboratory. He found that the composition of the naturally occurring compound was the same as the corresponding compound he prepared in his lab. This result became known as the Law of Constant Composition. This law states that no matter how or where a given compound is prepared or where it is found, it will always have the same composition. In order to use this law, it is necessary to define composition. According to Lavoisier, compounds are made up of elements, which can't be further broken down. Therefore, if one measures the total mass of the compound, breaks the compound down into the elements it is made up of and measures the mass of each element one can define the percent mass of any element say X as

$$\%X = \frac{m_X}{m_{comp}}(100) \quad . \tag{3}$$

Using Eq. (3) as the definition of composition one now has that for a given compound, all samples must have the same percent mass for each element it is made of. This argument can be turned around to define a compound. A compound is matter that satisfies the Law of Constant Composition. Note that not all matter (metal alloys or solutions) is a compound.

As an example, magnesium burns in oxygen to form a salt. When 5.0g of Mg is burned the salt weighs 8.3g. Determine the composition of this compound. The reaction can be written as

$$\text{Mg} + \text{O} \longrightarrow \text{Salt}$$
$$\text{Lab} \quad 5.0\text{g} \quad\quad m_O \quad\quad\quad 8.3\text{g}$$

and

$$m_O = 8.3\text{g} - 5.0\text{g} = 3.3\text{g} \ .$$

Then

$$\%\text{Mg} = \frac{5.0\text{g}}{8.3\text{g}}(100) = 60.\%$$

and

$$\%\text{O} = \frac{3.3\text{g}}{8.3\text{g}}(100) = 40.\% \ .$$

Note that the sum of the mass percents is one hundred as it should be.

EXERCISE

When 8.78g of iron (Fe) react with excess sulfur (S) 16.3g of a salt is produced. Determine the percent masses for iron and sulfur in this compound.

Dalton's Theory

John Dalton developed his atomic theory in 1803. It was published by Thomas Thompson in 1807 in the textbook "System of Chemistry" and by Dalton in 1808 in his book "A New System of Chemical Philosophy". His original version, found in his notebook, contained nine postulates. Below is a shortened version of most of his theory that uses modern chemical nomenclature.

1) All elements are made up of minute indivisible particles called atoms.

2) Atoms can neither be created nor destroyed.

3) All atoms of a given element have identical mass and other properties.

4) Atoms of different elements have different masses.

5) A particle of a compound is made up of a fixed whole number of atoms of its component elements. For example, if the compound is made of elements A and B it can be AB, AB_2, A_2B, etc. but never $AB_{1.5}$ or any other non-integer numbers.

One sees that postulates 2 (atoms are indestructible) and 3 (atoms have mass) explains the Law of Conservation of Mass. For the Law of Constant Composition consider the compound CO_2. Each CO_2 consists of one atom of C and two atoms of O. From postulate 3 all C atoms have the same mass (m_C) and all O atoms have the same mass (m_O). Therefore in each CO_2 the mass of C is m_C, the mass of O is $2m_O$ and the mass of CO_2 is $m_{CO_2} = m_C + 2m_O$. All molecules have the same percent masses of

$$\%C = \frac{m_C}{m_{CO_2}}(100) \quad \text{and} \quad \%O = \frac{2m_O}{m_{CO_2}}(100) \ .$$

Laboratory masses would require a huge number of CO_2's say N which would give a mass of CO_2 of Nm_{CO_2}, a mass of C of Nm_C and a mass of O of $2Nm_O$. Therefore the percent masses in the laboratory are

$$\%C = \frac{Nm_C}{Nm_{CO_2}}(100) = \frac{m_C}{m_{CO_2}}(100) \quad \text{and} \quad \%O = \frac{2Nm_O}{Nm_{CO_2}}(100) = \frac{2m_O}{m_{CO_2}}(100)$$

Which are always the same for CO_2. This can be done for any compound, which explains the Law of Constant Composition. One also notices that postulate 5 gives a new and very informative language for describing compounds.

One of the primary advantages of a theory is that it can predict new laws or experiments. Consider two compounds made up of the same elements say A and B. Suppose that the compounds are AB_2 and A_2B_3. If these compounds are compared for a fixed number of A atoms, say 2 ($2AB_2$ and A_2B_3) then the ratio of B atoms in the two compounds is four to three (4/3) or small whole numbers. Multiplying both 4 and 3 by m_B, the atomic mass of B, indicates that the ratio of the masses of B is also small whole numbers. Laboratory masses would require a huge number of A atoms

Chapter 1: The Experimental Laws | 5

say 2N and the total mass of A would be $2Nm_A$ in both compounds. The total mass of B in $2NAB_2$ would be $4Nm_B$ and in NA_2B_3 $3Nm_B$ and again the ratio of the masses of B would be 4/3. The general result is referred to as the Law of Multiple Proportions which states if more that one compound is made up of the same elements, for a fixed mass of one of the elements, the ratio of the masses in any two compounds for all other elements can be written in terms of small whole numbers.

To see how one can show that this law is satisfied with laboratory measurements consider three compounds that contain only nitrogen and oxygen. The first compound has %N=46.7 and %O=53.3. The second compound has %N=30.4 and %O=69.6. The third compound has %N=25.9 and %O=74.1. Converting percents into masses one can express this information in terms of chemical changes in the following form.

$$(NO)_1 \rightarrow N + O$$
Lab   100.g   46.7g   53.3g

$$(NO)_2 \rightarrow N + O$$
Lab   100.g   30.4g   69.6g

$$(NO)_3 \rightarrow N + O$$
Lab   100.g   25.9g   74.1g

However, to show that this law is satisfied, one must have the same mass for one of the elements, which is not the case above. If the N mass is fixed at 46.7g, then for the second compound one has

$$(NO)_2 \rightarrow N + O$$
Lab $m_{(NO)_2}$ 46.7g m_O

and using the percent mass of N
$$30.4 = \frac{46.7g}{m_{(NO)_2}}(100)$$

gives $m_{(NO)_2}$ =154g. Either the Law of Conservation of Mass or %O gives m_O=106.9g. A simpler approach is to multiply everything on the lab line in the second reaction by 46.7/30.4 and everything on the lab line in the third reaction by 46.7/25.9. Doing this gives the results below.

$$(NO)_1 \rightarrow N + O$$
Lab 100.g   46.7g   53.3g

$$(NO)_2 \rightarrow N + O$$
Lab 154.g   46.7g   106.9g

$$(NO)_3 \rightarrow N + O$$
Lab 180.g   46.7g   133.6g

From this one finds that the ratio of the masses of the O in the first two compounds is 106.9/53.3=2/1. The ratio for the first and third compounds is 133.6/53.3=2.5=5/2. Finally the ratio for the second and third compounds is 133.6/106.9=1.25=5/4. Therefore it is shown that these compounds all satisfy this law.

EXERCISE

Three compounds contain only the elements C, H and O. The first has %C=52.14%, %H=13.13% and %O=34.73%. The second has %C=38.70%, %H=9.744% and %O=51.56%. The third has %C=40.00%, %H=6.714 and %O=53.29. Show that these compounds satisfy the Law of Multiple Proportions.

If one considers the two questions in the introduction, "What are the building blocks?" and "What holds the building blocks together?", one notices that Dalton's Theory answers only the first question. However, consider elements. It is known that, under normal laboratory conditions, they can be solids such as iron, liquids such a mercury or gases such as hydrogen. For the solid or liquid cases, it is immaterial how many atoms are present in that dividing a gram in half still gives the same element. It therefore seems that atoms of some elements stay together to form liquids or solids and should therefore be attracted to each other. On the other hand, it seems that atoms of gaseous elements repulse each other in that if they were attracted to each other they should form a liquid or a solid. One would therefore expect that, based on these simple arguments, gaseous elements should be atoms of the elements. Dalton also felt that gaseous elements had to be atoms of the element.

The usefulness of this theory would be greatly enhanced if the atomic masses were known but there was no way to measure the mass of a single atom. It was known that the composition of water was 88% oxygen and 11% hydrogen. 100.g of water contain a very large number (N) of water molecules. However, the ratio of the mass of O to the mass of H is 88/11. For a chemical formula of the form H_xO_y, one must have that the mass of O is Nym_O and the mass of H is Nxm_H. Therefore one has

$$\frac{Nym_O}{Nxm_H} = \frac{ym_O}{xm_H} = \frac{88}{11} = 8 \tag{4}$$

and without knowing x and y (i.e., the chemical formula) the ratio of the atomic masses is not determined. To circumvent this difficulty, Dalton used an additional postulate he called the "rule of greatest simplicity". This rule states that if only one compound of two atoms A and B is known, unless there is some reason to the contrary, it is AB. If there is more than one compound, one is AB, the next is AB_2 or A_2B and so on. Since there was only one compound of H and O known, water, this rule said that x=y=1 (HO) so that

$$m_O = 8m_H$$

or that the oxygen atom is eight times heavier than an H atom. Dalton used this rule to establish a table of relative atomic masses. Others also reported relative atomic masses and they did not always agree. Of course the problem is that one can't directly observe the number of atoms in a compound so that a theory (good guess) is needed.

It was known at the time that when hydrogen reacted with oxygen to produce water at constant temperature and pressure, the ratio of the volume of H to the volume of O was two to one. In 1808, Gay-Lussac presented the results of his studies of other gas reactions at constant temperature and pressure, which showed the ratios of reactant, and product volumes were given in terms of small whole numbers. These results were called the Law of Combining Volumes. This law applies only to gaseous reactions. Gay-Lussac thought that these results suggest that equal volumes of gases with the same temperature and pressure contain an equal number of gas particles. Using a unit of volume of 1L, some examples of this law are

$$\begin{array}{cccc} & H & + \quad O & \rightarrow \quad (HO) \\ \text{Lab} & 2L & 1L & 2L \end{array}$$

$$\text{H} + \text{N} \rightarrow (\text{NH})$$
Lab 3L   1L   2L

and

$$\text{N} + \text{O} \rightarrow (\text{NO})$$
Lab 1L   1L   2L

where H, O and N are the symbols for the element gases (not chemical formulas) and the parenthesis indicate a compound of the elements inside them. It was difficult to reconcile this with Dalton's theory and Dalton did not accept the validity of this explanation. Dalton thought that gaseous elements must be atoms. For example if Gay-Lussac interpretation for the last reaction was correct, one would have only one N atom to make two identical molecules which leads to the unacceptable result of half a N in each molecule. If one accepted the equal volume, equal number hypothesis for gaseous elements but not for gaseous compounds, the H and O reaction would give a chemical formula for water of H_2O, the H and N reaction would give ammonia to be NH_3 and the N and O reaction NO. It was still difficult to understand why the volume of NO would be 2L instead of 1L.

In 1811, Amadeo Avogadro showed in a very simple way that the dilemma between Dalton's theory and Gay-Lussac interpretation could be resolved. In modern language, he proposed that gaseous elements could be if necessary diatomic molecules. With this assumption it was possible to show that all gases obeyed the equal volume, equal number hypothesis and no molecules with fractional atoms would be created.

When analyzing the experimental results of the Law of Combining Volumes it is helpful to list in order of use what the basic ideas are.

i. Gay-Lussac. Equal volumes contain an equal number of gas particles.

ii. Dalton. Gaseous elements should be atoms (starting point to get chemical formulas of molecules).

iii. Dalton. Molecules must contain whole atoms. Identify which molecules obtained in ii) have fractions of atoms.

iv. Avogadro. If necessary, gaseous elements can be diatomic molecules. Remove violations by making elements that have fractions into diatomic molecules. Note that the element is either an atom or a diatomic molecule and must be the same in all reactions. Then redo the chemical formulas for the product molecules.

For the experimental results above let one liter contain one gas particle so that Gay-Lussac's idea would become

$$H + O \rightarrow (HO)$$
number 2 1 2

$$H + N \rightarrow (NH)$$
number 3 1 2

and

$$N + O \rightarrow (NO)$$
number 1 1 2

where the number line gives the number of identical gas particles. One can now assume that the gas elements are atoms (idea ii)) and determine the corresponding chemical formulas of the product molecules as is shown below.

$$H + O \rightarrow HO_{\frac{1}{2}}$$
number 2 1 2

$$H + N \rightarrow N_{\frac{1}{2}}H_{\frac{3}{2}}$$
number 3 1 2

and

$$N + O \rightarrow N_{\frac{1}{2}}O_{\frac{1}{2}}$$
number 1 1 2

One sees that in each case the use of the second idea leads to a violation of the third idea. Therefore oxygen in the first reaction must be using the fourth idea a diatomic molecule, which also makes it a diatomic molecule in the third reaction. Using the fourth idea in the second reaction leads to both hydrogen and nitrogen being diatomic molecules. Therefore hydrogen in the first reaction is also a diatomic molecule and nitrogen in the third reaction is a diatomic molecule. The final results are given on the next page.

$$H_2 + O_2 \rightarrow H_2O$$
number 2   1   2

$$H_2 + N_2 \rightarrow NH_3$$
number 3   1   2

and

$$N_2 + O_2 \rightarrow NO$$
number 1   1   2

Notice that each chemical reaction now satisfies Dalton's theory in that the same number of atoms for each element is present on both sides of the reaction (no atoms are created or destroyed) and molecules have whole numbers of atoms. Using the correct formula for water in Eq. (4) (y=1 and x=2) gives

$$m_O = 16 m_H \quad .$$

The composition of ammonia is 82% N and 18% H. Using the same arguments as for water, one finds that

$$\frac{m_N}{3m_H} = \frac{82}{18} \quad \text{or} \quad m_N = 14 m_H \quad .$$

Finally the percent masses for NO are %N=46.7% and %O=53.3%. Therefore

$$\frac{m_N}{m_O} = \frac{46.7}{53.5} \quad \text{or} \quad m_N = 0.895 m_O = 0.895(16 m_H) = 14 m_H$$

Through these types of analysis as well as some others, the relative atomic masses of the growing number of elements were established and Dalton's Theory was accepted by virtually the entire scientific community.

EXERCISES

The following gas reactions with the volumes used and produced below them which are carried out at constant pressure and temperature involve three gaseous elements with chemical symbols of A, B and C. The parenthesis denote a compound and the element symbols inside indicate what elements are in the

Chapter 1: The Experimental Laws | 11

compound, e.g., (AB) is a compound made up of A and B. Use the four ideas on pg. 8 to determine the chemical formulas of the elements and the compounds.

$$A + B \rightarrow (AB)$$
Lab   4L   1L   1L

$$A + C \rightarrow (AC)$$
Lab   2L   1L   2L

$$B + C \rightarrow (BC)$$
Lab   1L   2L   2L

The compound in the first reaction has %A=14.3% and %B=85.7%. For the second compound %A=5.0% and %C=95%. For the third compound %B=24% and %C=76%. Determine the masses of B and C relative to the mass of A.

The following gas reactions with the volumes used and produced below them, which are carried out at constant pressure and temperature, involve three gaseous elements with chemical symbols of A, B and C. The parenthesis denote a compound and the element symbols inside indicate what elements are in the compound, e.g., (AB) is a compound made up of A and B. Use the four ideas on pg. 8 to determine the chemical formulas of the elements and the compounds.

$$A + B \rightarrow (AB)$$
Lab   3L   1L   1L

$$A + C \rightarrow (AC)$$
Lab   2L   1L   2L

$$B + C \rightarrow (BC)$$
Lab   1L   2L   1L

The compound in the first reaction has %A=21.4% and %B=78.6%. For the second compound %A=5.88% and %C=94.12%. For the third compound %B=14.67% and %C=85.33%. Determine the masses of B and C relative to the mass of A.

The Significance of Relative Atomic Masses

In the previous section it was found that the mass of an oxygen atom was sixteen times the mass of a hydrogen atom. Furthermore the element hydrogen consisted of H_2 molecules, the element oxygen consisted of O_2 molecules and water consisted of H_2O molecules. The reaction can be written in a way that has the number of atoms on both the reactant and product side the same for all elements (balanced). This is given by

$$2H_2 + O_2 \longrightarrow 2H_2O \quad .$$

Using a balanced reaction allows one to create very useful logical relations based on the fact that one mole of reaction involves two moles of H_2 and one mole of O_2 and two moles of water. This is summarized as

$$1 \text{molRx} = 2 \text{molH}_2 = 1 \text{molO}_2 = 2 \text{molH}_2\text{O}.$$

If the moles of H_2 is known the moles of O_2 would be

$$n_{O_2} = n_{H_2} \left(\frac{1 \text{molO}_2}{2 \text{molH}_2} \right)$$

and the moles of water would be

$$n_{H_2O} = n_{H_2} \left(\frac{2 \text{molH}_2O}{2 \text{molH}_2} \right) \quad .$$

This emphasizes why numbers of something are so important in Chemistry. It is because the conversion factors are ratios of small whole numbers.

One can't measure the mass of a single atom or molecule on a scale. It is only possible to weigh a very large number of atoms or molecules on a scale. Since the mass of a single hydrogen atom was not known at the time of this work (nineteenth century) it was not possible to determine how many hydrogen atoms would be present in one gram of hydrogen. However this number does exist and one can call this number one mole (mol) of hydrogen atoms. In other words one mole of Hydrogen atoms has a mass of one gram, which gives the logical relation

$$1 \text{molH} = 1\text{g} \quad .$$

Similarly H_2 has two H atoms so that

$$1 \text{molH}_2 = 2\text{g} \quad .$$

Also since an oxygen atom has a mass sixteen times that of a hydrogen atom, a mole of oxygen atoms would have a mass of sixteen times the mass of a mole of hydrogen atoms or

$$1 \text{molO} = 16\text{g} \quad .$$

Also since O_2 has two O atoms so that

$$1 \text{molO}_2 = 32\text{g} \quad .$$

Finally a water molecule consists of two H atoms and one O atom so that the mass of a water molecule would be eighteen times the mass of an H atom. Therefore the mass of a mole of water molecules would be eighteen times the mass of a mole of H atoms or

$$1 \text{molH}_2\text{O} = 18\text{g} \quad .$$

From these types of logical expressions we can form a very important conversion factor called the molar mass of something (M), which is the mass per mole of the something. For hydrogen one would have

$$M_H = \frac{1\text{g}}{\text{molH}} \quad ,$$

for H₂ it would be

$$M_{H_2} = \frac{2g}{\text{molH}_2},$$

for O it would be

$$M_O = \frac{16g}{\text{molO}},$$

for O₂ it would be

$$M_{O_2} = \frac{32g}{\text{molO}_2}$$

and for water it is

$$M_{H_2O} = \frac{18g}{\text{molH}_2O}.$$

In general the molar mass gives two conversion factors. If one has mass, multiplication by $1/M$ gives moles or

$$n(\text{mol}) = m(g)\left(\frac{1}{M\left(\frac{g}{\text{mol}}\right)}\right).$$

If one has moles, multiplication by M gives mass or

$$m(g) = n(\text{mol})\left(M\left(\frac{g}{\text{mol}}\right)\right).$$

These operations, converting mass into moles and moles into mass are very important in Chemistry.

Suppose now that one had 11g of hydrogen and you wanted to predict what mass of oxygen is needed and what mass of water would be produced in the reaction above. This can be set up in the following way.

$$2H_2 + O_2 \longrightarrow 2H_2O$$

lab   11g   m =   m =

mol

Using the molar masses given above and the integers in the reaction gives the following result.

$$2H_2 + O_2 \longrightarrow 2H_2O$$

lab   11g   m = 88g   m = 99g

$\downarrow \frac{1}{M_{H_2}}$ $\uparrow M_{O_2}$ $\uparrow M_{H_2O}$

mol $5.5\,mol\,H_2$ $\xrightarrow{\frac{1\,mol\,O_2}{2\,mol\,H_2}}$ $2.75\,mol\,O_2$ $\xrightarrow{\frac{2\,mol\,H_2O}{1\,mol\,O_2}}$ $5.5\,mol\,H_2O$

This analysis was based on the percent mass of H in water which was the information used in the Law of Constant Composition. One sees that the use of molar masses allows one to go to the "mole world" which corresponds to the mol line above where going between reactants and products is done by using the stoichiometric coefficients in the reaction. It is far easier to think in this world than in the lab world. Notice also that the actual number of atoms or molecules in a mole was not needed to solve the problem.

CHAPTER 2
THE AVERAGE ATOMIC MASS, THE MOLAR MASS AND EMPIRICAL AND MOLECULAR FORMULAS

Introduction

By the end of the nineteenth century Dalton's Theory was largely accepted and there were around ninety different elements known. The physicist were however somewhat skeptical that there would be over ninety different building blocks and set out to see if there were more basic building blocks than atoms. This led to several important discoveries that will be briefly mentioned here.

In 1896 Becquerel set out to determine whether phosphorescent substances emitted X-rays. On a cloudy day he placed some uranium ore on a photographic plate and put it in a drawer. When he removed it he found an exposure on the photographic plate. Since there was no light he discovered a spontaneous process called radioactivity. He found that the amount of radioactivity was proportional to the mass of the uranium. He also found that an ore called pitchblende gave more radioactivity than could be accounted for by the mass of uranium and gave the problem of finding it to Marie Curie. She and her husband Pierre Curie found a new element, polonium, which was hundreds of times more radioactive that uranium. They also found another element, radium, which was hundreds of thousands of times more radioactive than uranium and she was awarded two Nobel prizes.

In 1898 Rutherford found that two kinds of radiation was emitted in radioactivity he called alpha and beta rays. Beta rays were very penetrating but alpha rays could be stopped with a thin sheet of metal. Becquerel found that the beta rays had a negative charge and Rutherford showed that the alpha rays were positively charged. He also found that when an alpha ray was neutralized its spectrum coincided with the spectrum of a helium atom so that the alpha ray was a helium plus two ion. In 1903 Villard discovered a third uncharged ray that was more penetrating than beta rays and X-rays and Rutherford called them gamma rays. Therefore when a radioactive atom decays it can emit an alpha ray (a helium ion), a negative beta ray or a gamma ray.

A number of results were obtained for determining the radioactive decay series for various elements. In 1902 Rutherford and Soddy proposed that radioactive elements were undergoing spontaneous transformation into new elements. Thus the statement in Dalton's Theory that atoms can't be created or destroyed does not hold for radioactive elements. Rutherford also found that the half-life for the radioactive decay of uranium (the time for a beginning mass of uranium to decrease by an amount of one half) was about six billion years. Radium had a half-life of about seventeen hundred years. This emphasizes the problem with storing radioactive wastes. They must be stored "forever" in a very secure place. Fajans and Soddy generalized the proposal by Rutherford and Soddy in 1913 with the group displacement laws. These laws state that when an element decays and a beta ray is emitted the element produced lies one space to the right on the Periodic Table. When an element decays and an alpha ray is emitted the element produced lies two space to the left on the Periodic Table. Therefore if Uranium decays and emits a beta ray neptunium is produced. If Uranium decays and emits an alpha ray thorium is produced.

A Crookes tube is a device that consists of an evacuated glass tube with a cathode and an anode inside and connected to a high voltage source. This results in a negatively charged ray going from the anode to the cathode called a cathode ray. In 1880 Thomson performed numerous studies of cathode rays. He found that various kinds of electrodes produced the same cathode rays. He found that the ratio of the absolute value of the charge to the mass for all cathode rays was the same and had a value of 1.76×10^8 C/g. Since electrodes are made up of atoms, the cathode rays must come from atoms and since they come from various atoms they must be basic building blocks of an atom. He therefore referred to them as electrons and all atoms are made up of electrons. Therefore the concept of atoms being the most fundamental building blocks is not in actuality correct.

From 1906 to 1914 Millikan performed his famous oil drop experiments. In these experiments he used a chamber with the shape of a cylinder. In the top part of the cylinder was a positive plate and in the bottom was a negative plate and the voltage between the plates could be varied. Connected to the side of the chamber was a radioactive element that produced a culminated beam of alpha rays. The positive plate had a small hole in it and above it was a region that had an atomizer that produced uniform droplets of oil. Some of these drops would fall through the hole. When the alpha rays interacted with the oil drops it could produce charged oil drops. A negatively charged oil drop would slow down and a positively charged oil drop would speed up. Also on the side of the chamber between the plates was a telescope that allowed him to see an oil drop that fell through the hole. With the electric field turned off he could observe the time it took for the drops to fall a certain distance. He then observed the time it took for the charged drops to fall the same distance with the electric field turned on. He found that there was a minimum time change, a time change double this, a time change triple this and so on. He associated this with being due to one, two and three electrons. From this he established that the charge of the electron was -1.602×10^{-19} C. The charge to mass ratio gave the mass of the electron to be 9.109×10^{-28} g. This firmly established that the electron was a fundamental particle present in all atoms.

Using the electron as the fundamental particle Thomson created a model of the atom. The lightest atom hydrogen had a mass of about two thousand electrons. The hydrogen atom is also neutral so that a positive countercharge due to two thousand electrons was also needed. His model then consisted of a sphere of positive charge that was large enough to cancel the charges of the electrons. Inside this sphere were rings containing thousands of electrons. This model could be shown to be stable and not emit radiation.

In 1906 Rutherford sought out to probe an atom. What he needed was a massive, very small probe (smaller than an atom) that was moving at high speeds (an atomic bullet). The alpha ray had a mass of about four thousand electrons and a size of a helium ion. It also traveled with a speed about seven percent the speed of light. In essence these were atomic missiles. Alpha rays could penetrate a very thin sheet of metal. He used gold since it could be made very thin. His experiment consisted of using a radioactive source that could produce a culminated beam of alpha rays. A thin sheet of gold foil was placed in the path of the alpha rays. Also included were photographic plates lying on a circle whose diameter coincided with the path of the alpha rays. One was on the other side of the foil in the path of the alpha rays, some were near by and some were in back of the foil. Since the atom consisted of very light electrons he expected the alpha rays to pass through un-deflected (somewhat like firing a bullet through a cloud of feathers). When he did the experiment he found as expected that most of the alpha rays passed through un-deflected. However he also found that a few experienced minor deflections. He also found that on rare occasion an alpha ray would be deflected backwards. He later remarked that "It was about as credible as if you had fired a 15-inch shell at a piece of tissue paper and it came back and hit you". He could only account for these results by rejecting the Thomson model and constructing a new model. He concluded that practically all of the mass and positive charge was concentrated in a very small region of space he called the nucleus. A collision with this nucleus with a large mass would give large deflections but this would rarely happen because the nucleus is so small. The minor deflections were due to the positive charge of the alpha ray being repelled by the large positive charge of the nucleus. The electrons are outside the nucleus and account for the size of the atom. Since most of the volume is due to electrons the alpha rays mainly interact with electrons and are not deflected. An atom has a radius of around 10^{-10}m whereas the radius of a nucleus is around 10^{-15}m. If a baseball was the nucleus the atom would have a radius of about three miles.

Rutherford also set out to find if there was a fundamental unit of positive charge as was the case for negative charge (an electron). He used a tube with only nitrogen gas in it and passed alpha rays through it. After this he found the presence of hydrogen atoms, which had to be due to the interaction of alpha rays with nitrogen.

He obtained the same results for other gases and concluded that hydrogen nuclei were present in all atoms. This gives another fundamental building block he called the proton. The proton has a charge opposite to the electron, 1.602×10^{-19}C, and a mass 1837 times the mass of the electron 1.673×10^{-24}g. This led to a model for the nucleus to be enough protons to give the proper mass plus enough electrons to give the proper charge. A helium nucleus would be four protons and two electrons. The emission of a beta ray would be the loss of an electron in the nucleus.

In 1932 Joliot-Curie and Joliot discovered another fundamental building block that was neutral and slightly heavier that the proton (1.657×10^{-24}g) called a neutron. The model of a nucleus then became enough protons to give the correct charge and enough neutrons to give the correct mass. A helium nucleus is now two protons and two neutrons. The emission of a beta ray would be due to a neutron decaying into a proton and an electron ($n \rightarrow p + e^-$).

Using the group displacement laws it was found that the decay series for uranium and thorium ended with the element lead but the laws also predicted different masses for the lead atom and Soddy referred to these as isotopes of the element lead. Isotopes for the other nuclei in the decay series were also predicted. Theodore Richards and Max Lembert experimentally confirmed isotopes in 1914 when they determined that common lead had an atomic mass of about 207 times that of hydrogen but that lead from a uranium mineral was about 206 times that of hydrogen. Later on Honigschmid found that the atomic mass of lead from a thorium mineral was about 208 times that of hydrogen. Thomson was also able to show that lighter elements had isotopes. Thus the statement in Dalton's Theory that all atoms of a given element are identical is not correct.

Most elements have at least two stable isotopes and, in order to distinguish the isotopes, the notation

$$^A_Z X$$

is used where X is the symbol for the element, Z is the atomic number, which is the number of protons in the nucleus of the element, and A is the mass number, which is the number of protons plus the number of neutrons. All isotopes of an element have the same X and Z but A varies as the number of neutrons varies. For example, there are three isotopes for hydrogen given as

Hydrogen $\quad ^1_1H \quad ^2_1H \quad ^3_1H \qquad$,

two isotopes for nitrogen given as

Nitrogen $\quad {}^{14}_{7}N \quad {}^{15}_{7}N$,

three isotopes for oxygen given as

Oxygen $\quad {}^{16}_{8}O \quad {}^{17}_{8}O \quad {}^{18}_{8}O$,

uranium has 15 isotopes of which the most stable is ${}^{238}_{92}U$ and lead has 29 isotopes of which ${}^{204}_{82}Pb$, ${}^{206}_{82}Pb$, ${}^{207}_{82}Pb$ and ${}^{208}_{82}Pb$ are stable and are the final nuclei in decay series such as the ones mentioned above.

In SI units the values of the charges and masses of the fundamental particles are very small and therefore atomic units are often used. Atomic units use a unit of charge of

$$1 = 1.602 \times 10^{-19} C \quad .$$

The atomic mass unit (amu) is based on the ${}^{12}_{6}C$ isotope of carbon, which has a mass of 1.993×10^{-23}g. The mass of this isotope is defined to be exactly 12.00amu so that

$$12.00 \text{amu} = 1.993 \times 10^{-23} \text{g}$$

or

$$1 \text{amu} = 1.661 \times 10^{-24} \text{g} \quad .$$

The fundamental particles in atomic units are given below.

	symbol	mass	charge
electron	e^-	0.000548amu	-1
proton	p	1.007amu	$+1$
neutron	n	1.008amu	0

With this background we are now ready to clarify certain parts of Dalton's Theory. In the next section we will be able to determine with the help of nature the average atomic mass, what it is on the Periodic Table and how it can be used. This section will also show how the chemical formula of a salt can be obtained. The next section will quantify the mole and define the molar mass and how it is used. This information is used in the next section to determine empirical formulas and molecular formulas and the chapter will end with a preliminary look at stoichiometry that clearly demonstrates why it is based on using balanced chemical reactions.

The Average Atomic Mass

As pointed out in the introduction not all statements in Dalton's theory are absolutely correct. It is true that mater is made up of atoms but all atoms are made up of protons, electrons and neutrons. The statement that atoms can't be created or destroyed is not in general correct. New atoms can be created in radioactive decay or when a nucleus interacts with an alpha ray. However most of the work in Chemistry does not deal with these cases and the indestructibility of atoms can be used. Finally the statement that all atoms of a given element have the same mass is incorrect since elements have isotopes with different masses. This is problematic for the following reason. If you have a mass of an element on a scale and you know that each atom has the same mass and the mass of the atom then the number of atoms times the mass of an atom must be the mass on the scale. Then dividing the mass on the scale by the mass of an atom gives the number of atoms. However this worked only because each atom had the same mass. It does not work when all masses are not the same.

In order to clarify the concepts needed to get the number of atoms consider the following problem. There are marbles on a scale and the mass (m) is 30g. If all the marbles have a mass of 2g then the number of marbles (N) times 2g must be m or N(2g)=30g. Rearranging gives N=15 and there are fifteen marbles on the scale. Now suppose that some marbles have a mass of 1g and some have a mass of 2g. Without additional information the number can't be determined. One does know that the number of marbles with a mass of 1g (N_1) times 1g plus the number of marbles with a mass of 2g (N_2) times 2g must be 30g or

$$N_1(1g) + N_2(2g) = 30g .$$

However there are two unknowns and only one equation so that more information is needed. Suppose that you also know that 1/3 of the marbles have a mass of 1g and 2/3 of the marbles have a mass of 2g. Then N_1 must be N/3 and N_2 must be 2N/3. Then

$$\frac{N}{3}(1g) + \frac{2N}{3}(2g) = 30g$$

and

$$N\left(\frac{1}{3}(1g) + \frac{2}{3}(2g)\right) = 30g .$$

The term inside the parenthesis has two terms that are the fraction of marbles with a given mass time the mass. This is therefore the average mass of a marble and it is (5/3)g. Then

$$N = \frac{30g}{\left(\frac{5}{3}g\right)} = 18 \quad .$$

Therefore if the average mass of a marble is known the number of marbles can be obtained by dividing the mass by the average mass. However in order to know the average mass one must know the fractions with a given mass. These same concepts are applicable to atoms.

It turns out that nature gives us a real break. For any naturally occurring element found in nature the fractions of atoms for the isotopes are always the same. Naturally occurring means that the element doesn't come from something like radioactive decay. In actuality the vast majority of matter is made of naturally occurring elements. Instead of fractions the distributions of the isotopes are given as percents and called percent abundances. The fractions that are needed (as in the problem above) are obtained by dividing the percent abundances by 100.

Consider the element lead. As mentioned in the introduction, lead has four stable isotopes. In any matter containing naturally occurring lead the percent abundances and masses of the isotopes are 1.48% of the atoms are $^{204}_{82}Pb$ with mass 203.93amu, 23.6% are $^{206}_{82}Pb$ with mass 205.97amu, 22.6% are $^{207}_{82}Pb$ with mass 206.98amu and 52.3% are $^{208}_{82}Pb$ with mass 207.98amu. The question is, should all of this information be included on the periodic table or, since the same percentages occur everywhere, would an average value of this information be more useful? Since most chemicals are made from naturally occurring sources, the average would be more useful.

To calculate this average mass of a lead atom, one has

$$Am = \sum_i f_i m_i \quad (1)$$

where Am is the average atomic mass in amu, $\sum_i$ means to sum over i where i labels the different isotopes, f_i is the fraction of atoms that are isotope i and m_i is the mass of isotope i. The percent abundances are related to the fractions, f_i, by

Chapter 2: The Average Atomic Mass | 23

$$\text{\%abundance of } i = f_i(100). \quad (2)$$

For example, if there are two isotopes 1 and 2 with masses m_1 and m_2 and with $f_1=f_2=.5$, the average atomic mass would be $Am=(m_1+m_2)/2$ as expected. For the isotopes of lead letting i be 1, 2, 3 and 4 for the isotopes with mass number 204, 206, 207 and 208, Eqns. (2) and (1) give the average atomic mass for lead to be

$$Am = (0.0148)(203.93\text{amu}) + (0.236)(205.97\text{amu})$$
$$+ (0.226)(206.98\text{amu}) + (0.523)(207.98\text{amu})$$
$$= 207.18\text{amu}$$

which agrees with the value of the mass number on the periodic table. In general the mass number of any element on the Periodic Table can be interpreted as the average atomic mass of an atom of the element with a mass unit of amu.

Consider any element of the Periodic Table say X. The average atomic mass is the average mass of one atom and we have the logical relation

$$Am_X = 1\text{atomX}$$

where X is again any element. This gives the conversion factors

$$\frac{Am_X}{1\text{atomX}} \quad \text{and} \quad \frac{1\text{atomX}}{Am_X} \quad .$$

Rather than work with two conversion factors we will define the atomic mass of X, AM_X, as

$$AM_X = \frac{Am_X}{1\text{atomX}} \quad . \quad (3)$$

The units of AM_X are amu/atomX and when it multiplies the number of atoms of X, N_X, it gives the mass of the atoms, m_X, in amu. Then

$$m_X(\text{amu}) = N_X(\text{atomX})\left(AM_X\left(\frac{\text{amu}}{\text{atomX}}\right)\right) \quad .$$

The units of $1/AM_X$ are atomX/amu and when it multiplies the mass of the atoms of X it gives the number of atoms of X. Then

$$N_X(\text{atomX}) = m_X(\text{amu}) \left(\frac{1}{AM_X \left(\frac{\text{amu}}{\text{atomX}} \right)} \right).$$

Note that one conversion factor AM_X does both jobs.

Now consider 1.000g of naturally occurring lead on a laboratory scale and determine the number of atoms on the scale. The fundamental concept is that the total mass (m_{tot}) is the sum of the masses for each isotope and for each isotope its mass is the number of atoms of the isotope (N_i) times the mass of an atom of that isotope (m_i) which is ($N_i m_i$). Then

$$m_{tot} = \sum_i N_i m_i.$$

If the fractions for the isotopes are known then $N_i = N_{tot} f_i$ and

$$m_{tot} = N_{tot} \sum_i f_i m_i = N_{tot} Am = N_{tot}(\text{atomPb}) AM_{Pb}.$$

Rearranging gives

$$N_{tot}(\text{atomPb}) = \frac{m_{tot}}{AM_{Pb}}.$$

One notices that all that is needed to determine the total number of atoms on a scale is the average atomic mass given on the Periodic Table. For the 1.000g of lead atoms being considered one has that

$$N_{tot} = \frac{1.000\,g}{207.18\,\frac{amu}{atomPb}\left(1.661\times10^{-24}\,\frac{g}{amu}\right)} = 2.906\times10^{21}\,atomPb\;.$$

The number of atoms for any isotope can be found by multiplying the total number by the fraction for the isotope and knowing the atomic mass of the isotope would give in an obvious way the mass on the scale due to the isotope.

Dalton's Theory can now be modified and used for naturally occurring matter. Instead of saying that all atoms of a given element have the same mass one could say that all atoms of a naturally occurring element can be considered to have the same average mass. A revised version of Dalton's Theory that includes the new information is given below.

Dalton's Theory

1) All elements are made up of minute particles called atoms.

2) In elements not undergoing radioactive decay (nuclear reactions), atoms are not created or destroyed.

3) In naturally occurring elements, all atoms of a given element can be considered to have the same mass which is called the average atomic mass (Am).

4) Atoms of different naturally occurring elements have different average atomic masses.

5) A particle of a compound is made up of a fixed whole number of atoms of its component elements. For example, if the compound is made of elements A and B it can be AB, AB_2, A_2B, etc. but never $AB_{1.5}$ or any other non-integer numbers.

EXERCISES

Lithium (Li) has two stable isotopes. The first isotope has three neutrons, a mass of 6.0151amu and a relative abundance of 7.42%. The second isotope has four neutrons, a mass of 7.0160amu and a relative abundance of 92.58%. Give the symbols for each isotope. Determine the average atomic mass in amu. Determine how many lithium atoms are on a scale if the scale reads 1.768g.

($^{6}_{3}Li$ $^{7}_{3}Li$ Am = 6.942 amu, N_{tot} = 1.533x10^{23} atoms)

Neon (Ne) has three stable isotopes. The first isotope has ten neutrons, a mass of 19.9924amu and a relative abundance of 90.92%. The second isotope has eleven neutrons, a mass of 20.9940amu and a relative abundance of 0.257%. The third isotope has twelve neutrons, a mass of 21.9914amu and a relative abundance of 8.82%. Give the symbols for each isotope. Determine the average atomic mass in amu. Determine how many neon atoms are on a scale if the scale reads 14.79g.

($^{20}_{10}Ne$ $^{21}_{10}Ne$ $^{22}_{10}Ne$ Am = 20.02 amu, N_{tot} = 4.414x10^{23} atoms)

Sulfur (S) has four stable isotopes. The first isotope has sixteen neutrons, a mass of 31.9721amu and a relative abundance of 95.01%. The second isotope has seventeen neutrons, a mass of 32.9715amu and a relative abundance of 0.760%. The third isotope has eighteen neutrons, a mass of 33.9679amu and a relative abundance of 4.22%. The fourth isotope has twenty neutrons, a mass of 35.9671amu and a relative abundance of 0.0140%. Give the symbols for each isotope. Determine the average atomic mass in amu. Determine how many sulfur atoms are on a scale if the scale reads 0.00249g.

($^{32}_{16}S$ $^{33}_{16}S$ $^{34}_{16}S$ $^{36}_{16}S$ Am = 32.07 amu, N_{tot} = 4.675x10^{19} atoms)

Before leaving this section it is useful to consider, with the knowledge at hand, how a chemical formula of a salt could be obtained. Therefore consider the following problem. A compound is made up of iron and oxygen. When 10.0g of this compound are decomposed into its elements, the oxygen escapes and 7.00g of iron is produced. We wish to determine the chemical formula of this compound. The strategy for solving this problem is the following. Since the mass of iron and oxygen (conservation of mass) is known, statement 3 in Dalton's Theory allows one to determine the number of iron atoms and oxygen atoms. Statement 2 says that these atoms must have come from the compound. One therefore needs to re-express these numbers of atoms into the form given in statement 5. In problem solving, it is helpful to have a format that is common to all related problems. The format must represent the chemical change that occurred. In this case, the chemical change is the decomposition of the compound (FeO) into the elements Fe and O. The notation (FeO) simply indicates a compound of Fe and O of which the chemical formula is not known. This chemical change can be written as

$$(FeO) \rightarrow Fe + O$$

which is the beginning of the format. The format should also include the information in the problem and place it in correspondence with the chemicals. It should also distinguish whether it is observed in the lab or it is a "number of" quantity that would be in what will be called the "mole world" or mole for short. The extension of the format to include this is shown below.

$$(FeO) \rightarrow Fe + O$$

Lab

Mole

The information can now be added to the format as

$$\text{(FeO)} \rightarrow \text{Fe} + \text{O}$$
$$\text{Lab} \quad 10.0\,\text{g} \quad\quad 7.00\,\text{g} \quad\quad 3.00\,\text{g}$$

Mole

where grams is in the lab world. Since the chemical formula of the compound is not known, one can't take its mass into the mole world. However, since the average atomic mass (AM) of Fe and O is known, the number of Fe and O atoms can be determined which corresponds to a "number of" quantity in the mole world. Schematically the flow of the problem would be

$$\text{(FeO)} \rightarrow \text{Fe} + \text{O}$$
$$\text{Lab} \quad 10.0\,\text{g} \quad\quad 7.00\,\text{g} \quad\quad 3.00\,\text{g}$$
$$\downarrow \quad\quad \downarrow$$
$$\text{Mole} \quad \text{Fe}_{N_{Fe}}\text{O}_{N_O} \leftarrow N_{Fe} \quad\quad N_O$$

where N_{Fe} and N_O are the number of Fe and O atoms and these are the numbers in the compound. The problem is finished when these numbers are converted into small integers. The arrows going from the lab to the mole world represent dividing by the AM. This can also be included in the format to give

$$\text{(FeO)} \rightarrow \text{Fe} + \text{O}$$
$$\text{Lab} \quad 10.0\,\text{g} \quad\quad 7.00\,\text{g} \quad\quad 3.00\,\text{g}$$
$$\xrightarrow{\frac{1}{AM_{Fe}}}\downarrow \quad \xrightarrow{\frac{1}{AM_O}}\downarrow$$
$$\text{Mole} \quad \text{Fe}_{N_{Fe}}\text{O}_{N_O} \leftarrow N_{Fe} \quad\quad N_O$$

Since

$$AM_{Fe} = \left(55.847 \frac{\text{amu}}{\text{atomFe}}\right)\left(1.661 \times 10^{-24} \frac{\text{g}}{\text{amu}}\right) = 9.274 \times 10^{-23} \frac{\text{g}}{\text{atomFe}}$$

and in an analogous way

$$AM_O = 2.657 \times 10^{-23} \frac{\text{g}}{\text{atomO}} \quad ,$$

using this in the format gives the result below.

$$\text{(FeO)} \rightarrow \text{Fe} + \text{O}$$

Lab   10.0g   7.00g   3.00g

$$\xrightarrow{\frac{1}{AM_{Fe}}} \downarrow \qquad \xrightarrow{\frac{1}{AM_{O}}} \downarrow$$

Mole $Fe_{7.55 \times 10^{22}} O_{1.13 \times 10^{23}}$ $\leftarrow$ $N_{Fe} = 7.55 \times 10^{22}$ $N_O = 1.13 \times 10^{23}$

The huge numbers of atoms involved are the actual numbers in the compound. However, the chemical formula is based on the smallest integers that give the relative number of atoms in any amount of the compound. Since the number of Fe atoms is smaller than the number of O atoms one has

$$\frac{N_O}{N_{Fe}} = \frac{1.13 \times 10^{23}}{7.55 \times 10^{22}} = 1.50$$

or

$$\frac{N_O}{N_{Fe}} = \frac{3}{2}$$

so that the chemical formula is Fe_2O_3.

The Mole Concept

In the previous section, the chemical formula of Fe_2O_3 was obtained using the fundamental concept that the number of atoms of an element could be determined by dividing the laboratory mass of the element by the AM of the element. This procedure, although quite straightforward, suffers from two major inconveniences. First, in order to cancel the mass unit of grams in the lab, the masses on the Periodic Table must be converted from amu to grams. Second, the numbers of atoms involved are about 10^{22} to 10^{23} and dealing with huge numbers is not very appealing. Since all lab measurements of mass on scales involves very large numbers, it would seem reasonable to introduce a new unit for a large number that would eliminate these large numbers (e.g. 1mile=6.336×10^4in).

In introducing this unit, it would be convenient to use the same number (the mass number) on the Periodic Table but with a mass unit of grams. This would get rid of the conversion factor to change mass units needed in the previous section. However with a mass number with units of grams one would not be considering a single atom but rather many atoms. Then one must know how many atoms would be present in this mass. For example, if one wanted to interpret the number for oxygen, 15.9994, with a mass unit of grams, the number of oxygen atoms (N) in this mass would be

$$N(\text{atomO})\left(AM_O\left(\frac{\text{amu}}{\text{atomO}}\right)\right) = 15.9994 \text{g}$$

or

$$N = \frac{15.9994 \text{g}}{\left(15.9994 \frac{\text{amu}}{\text{atomO}}\right)\left(1.6606 \times 10^{-24} \frac{\text{g}}{\text{amu}}\right)} = 6.022 \times 10^{23} \text{atomO}$$

which is a huge number. One notices that the number for the mass in grams cancels the number for the mass in amu/atomO and that this will be true for any element on the Periodic Table. Therefore, this is a unique number for all elements and is the number of atoms when the Periodic Table mass number has units of grams. This number is called Avogadro's number (N_A) and one has

$$N_A = 6.022 \times 10^{23} \quad .$$

One notices that the conversion from amu to grams is just Avogadro's number so that

$$1\text{g} = N_A \text{amu} \quad .$$

It should now be clear that there are two interpretations of the mass number for any element on the Periodic Table. One interpretation is that the number is the average mass of an atom of the element in units of amu/atom. The second interpretation is that it is the mass of Avogadro's number of atoms with units of grams per N_A number of atoms. It is convenient to introduce a number unit that is equal to Avogadro's number and is called a mole (mol). It is defined as

$$1\text{mol} = N_A = 6.022 \times 10^{23} \quad . \tag{2}$$

The second interpretation is now that the mass number for any element is equal to the mass in grams of one mol of that element. This can be used to create another conversion factor that is very useful. Consider again oxygen. One has that one mol of O is equal to 15.9994g of O and we have the logical relation

$$1\text{molO} = 15.9994 \text{g} \quad .$$

This gives two conversion factors

$$1 = \frac{15.9994 \text{g}}{1 \text{molO}} \quad \text{and} \quad 1 = \frac{1 \text{molO}}{15.9994 \text{g}} \quad .$$

Rather than working with two conversion factors it is convenient to define the molar mass (M_X) for any element (X) as the mass number of X (M_X) on the Periodic Table with units of grams per mole so that

$$M_X = M_X \left(\frac{\text{g}}{\text{molX}} \right) \tag{4}$$

and the conversion factors are now

$$1 = M_X = \frac{1}{M_X} \quad . \tag{5}$$

In the format, the lab world and mole world was introduced. In the lab world were masses and in the mole world there were "numbers of". The solution of the problem for the empirical formula involved going from the lab world to the mole world. Suppose that in the lab, there were 15.9994g of oxygen. Since one mol of O has a mass of 15.9994g, this mass must obviously correspond to one mole. Using the conversion factor in Eq.(5) one has

$$n_O(\text{mol}) = 15.9994 \text{g} \left(\frac{1}{M_O} \right) = 15.9994 \text{g} \left(\frac{1}{15.9994 \frac{\text{g}}{\text{mol}}} \right) = 1 \text{mol}$$

where the symbol n_X means the number of moles of X (O in this case). In more general terms one has

$$n_X = \left(\frac{1}{M_X} \right) m_X \tag{6}$$
$$\text{mole} \quad \leftarrow \quad \text{lab}$$

where m_X is the mass of X. The second line in Eq. (6) emphasizes that the mass (m_X) in the lab world is converted into the number of moles of X (n_X) in the mole world by the conversion factor ($\leftarrow$) of $1/M_X$. Rearranging Eq. (6) gives

$$m_X = (M_X) \; n_X \quad (7)$$
$$\text{lab} \quad \leftarrow \quad \text{mole}$$

where the second line in Eq. (7) emphasizes that the number of moles (n_X) in the mole world is converted into the mass (m_X) in the lab world by the conversion factor ($\leftarrow$) of M_X. Considering oxygen again, if there are 8.00g of O in the lab world, Eq. (6) gives

$$n_O = \left(\frac{1}{15.9994 \frac{g}{mol}} \right)(8.00\,g) = 0.500\,mol$$

which is the obvious result. If there are 2.00mol of O in the mole world, Eq. (7) gives

$$m_O = \left(15.9994 \frac{g}{mol} \right)(2.00\,mol) = 32.0\,g$$

which is again an obvious result. Equations (6) and (7) are very important expressions because they link the lab world and mole world together for masses in grams in the lab world.

Thus far, only elements have been considered. For the example in the introduction, consider a single unit of Fe_2O_3 that is broken down into atoms. This process can be written as

$$Fe_2O_3 \;\rightarrow\; 2Fe \;+\; 3O$$
$$m_{Fe_2O_3} \;=\; 2AM_{Fe} \;+\; 3AM_O$$

where the second line is based on the statements in Dalton's Theory and $m_{Fe_2O_3}$ is the mass of one Fe_2O_3 unit and all masses have mass units of grams. If one considers Avogadro's number of units then the first line becomes

$$1\,mol\,Fe_2O_3 \;\rightarrow\; 2\,mol\,Fe \;+\; 3\,mol\,O$$

and multiplying the second line by Avogadro's number of atoms divided by one mole gives

Chapter 2: The Average Atomic Mass | 33

$$M_{Fe_2O_3} = \left(\frac{2\text{molFe}}{1\text{molFe}_2O_3}\right)M_{Fe} + \left(\frac{3\text{molO}}{1\text{molFe}_2O_3}\right)M_O$$

where

$$M_X = AM_X\left(\frac{\text{g}}{\text{atomX}}\right)\left(\frac{N_A \text{atomX}}{1\text{molX}}\right)$$

was used. Since $M_{Fe_2O_3}$ which has units of g/mol is the mass of one mole of Fe_2O_3, it is the molar mass of Fe_2O_3. The important thing to realize is that the molar mass of Fe_2O_3 was obtained from the Periodic Table by adding together the molar masses of all the atoms that were present in the chemical formula. In more general terms, for a compound made up of elements X, Y, Z, ···· with a chemical formula of $X_xY_yZ_z$····, the molar mass of the compound is given by (units of compound conversion factors are suppressed)

$$M_{X_xY_yZ_z\cdots} = xM_X + yM_Y + zM_Z + \cdots \quad (7)$$

so that the molar mass of any compound with known chemical formula can be easily obtained by using the molar masses on the Periodic Table. For Fe_2O_3, one obtains

$$M_{Fe_2O_3} = 2\left(55.847\frac{\text{g}}{\text{mol}}\right) + 3\left(15.9994\frac{\text{g}}{\text{mol}}\right) = 159.692\frac{\text{g}}{\text{mol}}.$$

EXERCISES

Determine the number of moles for the following.

i) 10.75g of Fe (0.192mol)

ii) 1.42g of S (0.0443mol)

iii) 123g of H_2O (6.83mol)

iv) 41.5g of Na_2SO_4 (0.292mol)

Determine the mass in grams of the following.

    i)   1.25mol of Co (73.7g)

    ii)   0.467mol of Br (37.3g)

    iii)   3.58mol of $KClO_4$ (496g)

    iv) 5.28×10^{-4} mol of $C_{10}H_{10}O_4$ (0.0103g)

Consider again the iron oxygen problem treated in the previous section. Using molar masses one has

$$\begin{array}{ccccc}
 & (FeO) & \rightarrow & Fe & + & O \\
\text{Lab} & 10.0g & & 7.00g & & 3.00g \\
 & & & \downarrow \frac{1}{M_{Fe}} & & \downarrow \frac{1}{M_O} \\
\text{Mole} & Fe_{n_{Fe}}O_{n_O} & \leftarrow & n_{Fe} & & n_O
\end{array}$$

and using the molar mass for Fe and O gives

$$\begin{array}{ccccc}
 & (FeO) & \rightarrow & Fe & + & O \\
\text{Lab} & 10.0g & & 7.00g & & 3.00g \\
 & & & \downarrow \frac{1}{M_{Fe}} & & \downarrow \frac{1}{M_O} \\
\text{Mole} & Fe_{.1253}O_{.1875} & \leftarrow & n_{Fe}=.1253\text{mol} & & n_O=.1875\text{mol}
\end{array}$$

Then

$$\frac{n_O}{n_{Fe}} = \frac{.1875 \text{mol}}{.1253 \text{mol}} = 1.496 = \frac{3}{2}$$

so that the empirical formula is Fe_2O_3 which is the same result. This should not be surprising since all that was done was to introduce quantities, molar masses, for Avogadro's number of atoms instead of single atom quantities. The introduction of these molar quantities has resulted in two major advantages. First, the mass numbers on the Periodic Table can be used directly without converting mass units when interpreted in terms of grams per mole. Secondly, the numbers involved in the solution are much more easy to manage and interpret especially if one keeps in mind that mol is simply a unit for a large number (Avogadro's number).

For any compound there are logical relations based on the chemical formula. For example the salt Na_2SO_4 has

$$1\,mol\,Na_2SO_4 = 2\,mol\,Na = 1\,mol\,S = 4\,mol\,O$$

which give six logical relations and therefore twelve conversion factors. Rather than writing them all down it is much simpler to use the chemical formula to create the ones that are needed. For example if the moles of salt is known the moles of O would be obtained by multiplying the moles of salt by $4molO/1molNa_2SO_4$. If the mass of Na in the salt is known the mass of O is

$$m_O = m_{Na}\left(\frac{1}{M_{Na}}\right)\left(\frac{4\,mol\,O}{2\,mol\,Na}\right)M_O \quad .$$

One can also obtain the percent masses in a compound by using the molar mass of the compound. For K in K_2SO_4 the percent mass of K (%K) is

$$\%K = \frac{m_K}{m_{tot}}(100)$$

and considering one mole of the above compound,

$$m_{tot} = 1\,mol\,K_2SO_4\left(M_{K_2SO_4}\right)$$

and

$$m_K = 1\,mol\,K_2SO_4\left(\frac{2\,mol\,K}{1\,mol\,K_2SO_4}\right)(M_K)$$

so that

$$\%K = \frac{2\,mol\,K\left(M_K\right)}{1\,mol\,K_2SO_4\left(M_{K_2SO_4}\right)}(100)$$

with analogous equations for %S and %O.

As an example, consider the compound potassium sulfate, K_2SO_4 where the molar mass is 174.2602 g/mol. Then the percent masses are

$$\%K = \frac{2\,mol\,K\,M_K}{1\,mol\,K_2SO_4\,M_{K_2SO_4}}(100) = \frac{2(39.0983g)}{174.2602g}(100) = 44.87\%$$

$$\%S = \frac{1\,mol\,S\,M_S}{1\,mol\,K_2SO_4\,M_{K_2SO_4}}(100) = \frac{32.066g}{174.2602g}(100) = 18.40\%$$

and

$$\%O = 100 - \%K - \%S = 100 - 44.87 - 18.40 = 36.73\% \ .$$

EXERCISES

Determine the mass and the number of moles of the elements present in 6.45g of $Fe(ClO_4)_2$. (n_{Fe}=0.0253mol, m_{Fe}=1.41g, n_{Cl}=0.0506mol, m_{Cl}=1.69g, n_O=0.203mol, m_O=3.25g)

Determine the percent masses of the elements in $Mn_2(Cr_2O_7)_3$. (%Mn=14.5, %Cr=41.2)

Empirical Formulas

Empirical formulas are chemical formulas that are obtained from experimental information that gives only the lowest set of integers for the compound. For ionic compounds, this is the correct chemical formula but for molecular compounds it is not necessarily the molecular formula. Consider hydrogen peroxide which is made up of H and O. When 10.0g of this compound is decomposed, 9.41g of O and .593g of H are produced. Using the previous format gives

$$
\begin{array}{ccccccc}
& (HO) & \rightarrow & H & + & O \\
\text{lab} & 10.0\,g & & 0.593\,g & & 9.41\,g \\
& & & \downarrow \frac{1}{M_H} & & \downarrow \frac{1}{M_O} \\
\text{mole} & H_{.588}O_{.588} & \leftarrow & n_H = .588\,\text{mol} & & n_O = 0.588\,\text{mol}
\end{array}
$$

so that the empirical formula is HO. This gives only the lowest set of integers not necessarily the set of integers in the molecular formula. The molecular formula could be HO or H_2O_2 or H_3O_3 or any multiple of the empirical formula. However one knows that the molar mass of the molecule must be a multiple of the molar mass of the empirical formula. Therefore the number of empirical formulas in the molecular formula can be found by dividing the molar mass of the molecule by the molar mass of the empirical formula. The molar mass of hydrogen peroxide is 34.02g/mol and the molar mass of HO is 17.01g/mol so that

$$\frac{M_{H_2O_2}}{M_{HO}} = \frac{34.02\,\frac{g}{mol}}{17.01\,\frac{g}{mol}} = 2$$

and molecular formula is H_2O_2

Although this area of Chemistry, the determination of empirical formulas, molecular formulas, etc., seems to contain a variety of different kinds of problems, there are three features that are common to all problems. The first is that all problems can be described by a chemical change, i.e., a chemical reaction. The second feature is that all problems have measurements (numbers with units). The third feature is that all problems have an objective. These features can be used in solving problems by turning them into the ordered set of questions given below. Rather than reading a problem very carefully and asking a single question "Do I know how to solve it?", which leads to a dead end street if the answer is no, this procedure uses an answerable set of questions for all problems.

1) What is the Chemistry?

2) What are the measurements?

3) What is the objective?

A problem is read with the intent of answering each question in order. The answer to each question is placed into a format that is common to all problems so that all problems look the same. What this procedure actually does is it reformulates the problem into something that is familiar, the format, which can be used to go from the information to the objective.

In some problems, the information about the unknown compound is its mass and the masses of the elements it is made up of. If the compound is made up of X and Y where X and Y are arbitrary known elements, then the answers to questions 1 and 2 are given by

$$\begin{array}{cccccc} & (XY) & \to & X & + & Y \\ \text{lab} & m_{(XY)} & & m_X & & m_Y \end{array}.$$

mole

If instead the measurements are percent masses, one hundred grams of the compound can be used so that percent masses become masses in grams and

$$\begin{array}{cccccc} & (XY) & \to & X & + & Y \\ \text{lab} & 100.\text{g} & & \%X\text{g} & & \%Y\text{g} \end{array}.$$

mole

In either case the solution of the problem is given by

$$\begin{array}{cccccc} & (XY) & \to & X & + & Y \\ \text{lab} & m_{(XY)} & & m_X & & m_Y \\ & & & \frac{1}{M_X}\downarrow & & \frac{1}{M_Y}\downarrow \\ \text{mole} & X_{n_X}Y_{n_Y} & \leftarrow & n_X & & n_Y \end{array}$$

where the n_X and n_Y are converted into integers.

As an example, a compound containing cobalt and sulfur is 55.1% Co and 44.9% S by mass. Determine the empirical formula. Using the format gives

$$\text{lab} \quad \begin{array}{c} (CoS) \\ 100\,g \end{array} \rightarrow \begin{array}{c} Co \\ 55.1g \end{array} + \begin{array}{c} S \\ 44.9g \end{array}$$

$$\downarrow \tfrac{1}{M_{Co}} \qquad \downarrow \tfrac{1}{M_S}$$

$$\text{mole} \quad Co_{.935}S_{1.40} \leftarrow n_{Co} = .935\,\text{mol} \qquad n_S = 1.40\,\text{mol}$$

and since

$$\frac{1.40}{.935} = 1.49 = \frac{3}{2}$$

the empirical formula is Co_2S_3 which is also the chemical formula of a salt.

As a second example, a compound containing C, H and O has a molar mass of 90.08g/mol. A mass of 5.000g of this compound produced 2.00g of C, 0.336g of H and 2.66g of O. Determine the molecular formula. Using the format gives

$$\text{lab} \quad \begin{array}{c} (CHO) \\ 5.00g \end{array} \rightarrow \begin{array}{c} C \\ 2.00g \end{array} + \begin{array}{c} H \\ .336g \end{array} + \begin{array}{c} O \\ 2.66g \end{array}$$

$$\downarrow \tfrac{1}{M_C} \qquad \downarrow \tfrac{1}{M_H} \qquad \downarrow \tfrac{1}{M_O}$$

$$\text{mole} \quad C_{.166}H_{.333}O_{.166} \leftarrow n_C = .166 \qquad n_H = .333 \qquad n_O = .166$$

Dividing by .166 gives the empirical formula of CH_2O. The molar mass of the empirical formula is 30.03g/mol so that

$$\frac{90.08\,\tfrac{g}{mol}}{30.03\,\tfrac{g}{mol}} = 2.9997 = 3$$

and the molecular formula is $C_3H_6O_3$. In the above analysis, the units of mol in the mole world are no longer written since, in the mole world, the unit is almost always mol and is hence understood.

Finally, consider a metal oxide where the metal is unknown. When 5.00g is decomposed, 2.65g of the unknown metal and 2.35g of O are produced. Determine what metal(s) the unknown can be. Let M be the unknown metal and since the metal oxide is a compound, let the chemical formula be M_xO_y. Using the format gives

$$M_xO_y \rightarrow xM + yO$$

lab 5.00g   2.65g   2.35g

$$\text{mole} \qquad n_M = \tfrac{x}{y}(.148) \xleftarrow{\frac{x}{y}} n_O = .148$$

with $M_M \uparrow$ and $\frac{1}{M_O} \downarrow$ indicated above.

where

$$n_M = n_O\left(\frac{x\,\text{molM}}{y\,\text{molO}}\right).$$

If x and y are known, then the identity of the metal is determined by determining its molar mass and finding the molar mass on the Periodic Table. Then

$$M_M = \frac{m_M}{n_M} = \frac{y(2.65\text{g})}{xn_O} = \frac{y(2.65\text{g})}{x(.148\text{mol})} = \left(\frac{y}{x}\right)17.9\frac{\text{g}}{\text{mol}}.$$

Oxygen has a negative two charge and a metal can have a charge of +1, +2, +3, +4, +5, etc. Suppose that the metal has a charge of +1. Then x=2 and y=1 for charge neutrality and the molar mass of M is

$$M_M = \left(\frac{1}{2}\right)\left(17.9\frac{\text{g}}{\text{mol}}\right) = 8.95\frac{\text{g}}{\text{mol}}.$$

This is close to the molar mass of Be but Be is an alkaline earth metal that has a charge of +2 in ionic compounds so that this can be ruled out. If the metal has a charge of +2, x=y=1 and the molar mass of the metal is 17.9g/mol. This does not agree with the molar masses on the Periodic table so that this is also ruled out. If the charge is +3 then x=2 and y=3 and the molar mass is 26.9g/mol which agrees with the molar mass of Al (which as an ion has a +3 charge) on the Periodic Table. Therefore the compound could be Al_2O_3 which is aluminum oxide. If the charge is +4, x=1 and y=2 so that the molar mass would be 35.8g/mol. This is close to the one for chlorine but since Cl is a nonmetal, this is ruled out. For a +5 charge, x=2 and y=5 so that the molar mass is 44.8g/mol which is close to scandium. However, Sc can not have a +5 charge (it has only three highest energy electrons) so it is ruled out. Finally, for a +6 charge, x=1 and y=3 which gives a molar mass of 53.7g/mol which is not on the Periodic Table. Therefore the conclusion is that the unknown metal is Al and the chemical formula is Al_2O_3.

EXERCISES

A salt of Chromium (Cr) and oxygen (O) has %Cr of 68.4 and %O of 31.6. Determine the empirical formula. (Cr_2O_3)

Determine the empirical formula for a compound of Na and O if 2.74g of the compound produced 2.03g of Na. (Na_2O)

When 4.25g of a compound containing only C, H and O was decomposed into its elements, 1.64g of C, 0.414g of H and 2.19g of O were produced. The molar mass of this compound is 62.07g/mol. Determine the molecular formula for this compound. ($C_2H_6O_2$)

When 5.00g of a salt of an unknown metal and S decomposed, it produced 2.334g of S. Determine the identity of the metal. (Fe or Zr)

In some problems the unknown compound is broken down into usually known, simpler compounds. Often this is done by having the unknown undergo a reaction with a known reactant to form known products. An example of this is the combustion of an organic molecule that contains only C and H with excess oxygen to produce water and carbon dioxide. If the mass of the unknown and the masses of the water and carbon dioxide are given the format for the answer to questions 1 and 2 becomes

$$\text{(CH)} + \text{O}_2 \rightarrow \text{CO}_2 + \text{H}_2\text{O}$$

lab $\quad m_{(CH)} \quad\quad m_{O_2} \quad\quad m_{CO_2} \quad\quad m_{H_2O}$

mole

where the mass of O_2 is obtained from the conservation of mass. Since the molar masses of water and carbon dioxide are known and the unknown does not contain O, the general solution is

$$\text{(CH)} + \text{O}_2 \rightarrow \text{CO}_2 + \text{H}_2\text{O}$$

lab $\quad m_{(CH)} \quad\quad m_{O_2} \quad\quad m_{CO_2} \quad\quad m_{H_2O}$

$\quad\quad\quad\quad\quad\quad \frac{1}{M_{O_2}}\downarrow \quad \frac{1}{M_{CO_2}}\downarrow \quad \frac{1}{M_{H_2O}}\downarrow$

mole $\quad\quad\quad\quad\quad n_{O_2} \quad\quad n_{CO_2} \quad\quad n_{H_2O}$

$\quad\quad\quad\quad\quad\quad \frac{2}{1}\downarrow \quad \frac{1}{1}\downarrow\frac{2}{1} \quad \frac{2}{1}\downarrow\frac{1}{1}$

$\quad\quad C_{n_C}H_{n_H} \quad n_O \leftarrow n_C \quad n_O \quad\quad n_H \quad n_O$

and the n_C and n_H need to be converted to integers. Notice that the moles of carbon dioxide and water must be converted into moles of C and H, which are needed in the compound. The above analysis includes the determination of the moles of oxygen but, obviously, this is not needed since the unknown does not contain oxygen. However this is necessary when the unknown also includes oxygen. In this case, the format becomes

$$\text{(CHO)} + \text{O}_2 \rightarrow \text{CO}_2 + \text{H}_2\text{O}$$

lab $\quad m_{(CHO)} \quad\quad m_{O_2} \quad\quad m_{CO_2} \quad\quad m_{H_2O}$

$\quad\quad\quad\quad\quad\quad \frac{1}{M_{O_2}}\downarrow \quad \frac{1}{M_{CO_2}}\downarrow \quad \frac{1}{M_{H_2O}}\downarrow$

mole $\quad\quad\quad\quad\quad n_{O_2} \quad\quad n_{CO_2} \quad\quad n_{H_2O}$

$\quad\quad\quad\quad\quad\quad \frac{2}{1}\downarrow \quad \frac{1}{1}\downarrow\frac{2}{1} \quad \frac{2}{1}\downarrow\frac{1}{1}$

$\quad\quad C_{n_C}H_{n_H}O_{n_O} \quad n_{O(O_2)} \leftarrow n_C \quad n_{O(CO_2)} \quad n_H \quad n_{O(H_2O)}$

where the additional information in the subscripts for n_O indicate from which compound it came from. Since the number of O atoms on both sides of the reaction must be equal, n_O in the compound must be

$$n_O = n_{O(CO_2)} + n_{O(H_2O)} - n_{O(O_2)}$$

so that the empirical formula can be determined.

Another illustration of this are hydrated salts that when heated produce water and the anhydrous salt. The number of water molecules associated with hydrated salt is written in the form salt $\cdot$ xH$_2$O where x is usually an integer but sometimes a fraction. If the mass of the hydrated salt and the masses of the anhydrous salt and water are given, the answers to questions 1 and 2 give the format shown below.

$$\text{salt} \cdot xH_2O \rightarrow \text{salt} + xH_2O$$

lab $\quad m_{salt \cdot xH_2O} \quad\quad m_{salt} \quad\quad m_{H_2O}$

mole

The general solution is

$$\text{salt} \cdot xH_2O \rightarrow \text{salt} + xH_2O$$

lab $\quad m_{salt \cdot xH_2O} \quad\quad m_{salt} \quad\quad m_{H_2O}$

$$\frac{1}{M_{salt}} \downarrow \quad\quad \frac{1}{M_{H_2O}} \downarrow$$

mole $\quad\quad\quad\quad n_{salt} \xrightarrow{\frac{x}{1}} n_{H_2O}$

and since

$$n_{H_2O} = n_{salt}\left(\frac{x\, mol\, H_2O}{1\, mol\, salt}\right)$$

then

$$x = \frac{n_{H_2O}}{n_{salt}}$$

where the units have been canceled.

As an example, 7.50g of a compound containing C and H is combusted with excess oxygen and 22.0g of carbon dioxide and 13.5g of water are produced. The molar mass of the compound is 30.07g/mol. Determine the molecular formula. Using the format gives

$$\begin{array}{ccccccc}
& (CH) & + & O_2 & \rightarrow & CO_2 & + & H_2O \\
\text{lab} & 7.50g & & xs & & 22.0g & & 13.5 \\
& & & & & \downarrow \frac{1}{M_{CO_2}} & & \downarrow \frac{1}{M_{H_2O}} \\
\text{mole} & & & & & n_{CO_2}=.500 & & n_{H_2O}=.749 \\
& & & & & \downarrow \frac{1}{1} & & \downarrow \frac{2}{1} \\
& C_{.500}H_{1.50} & & & \leftarrow & n_C=.500 & & n_H=1.50
\end{array}$$

where the molar mass of CO_2 is 44.011g/mol, the molar mass of water is 18.016g/mol and, since the compound does not contain O, the analysis for O is not needed. The empirical formula is CH_3 which has a molar mass of 15.04g/mol so that 30.07/15.04=2. Therefore the molecular formula is C_2H_6.

As a second example, maleic acid contains C, H and O and has a molar mass of 116.08g/mol. When 10.0g is combusted with excess oxygen, 15.2g of carbon dioxide and 3.10g of water are produced. Determine the molecular formula. Using the format gives

$$\begin{array}{ccccccc}
& (CHO) & + & O_2 & \rightarrow & CO_2 & + & H_2O \\
\text{lab} & 10.0g & & 8.3g & & 15.2g & & 3.10g \\
& & & \downarrow \frac{1}{M_{O_2}} & & \downarrow \frac{1}{M_{CO_2}} & & \downarrow \frac{1}{M_{H_2O}} \\
\text{mole} & & & n_{O_2}=.259 & & n_{CO_2}=.345 & & n_{H_2O}=.172 \\
& & & \downarrow \frac{2}{1} & & \downarrow \frac{1}{1} \downarrow \frac{2}{1} & & \downarrow \frac{2}{1} \downarrow \frac{1}{1} \\
& & & & & n_C=.345 & & n_H=.344 \\
& C_{.345}H_{.345}O_{.345} & & n_O=.519 & \leftarrow & n_O=.690 & & n_O=.172
\end{array}$$

where the number of moles of O in the compound is

$$n_O = .692 + .172 - .519 = .345 \ .$$

This gives the empirical formula of CHO with a molar mass of 29.02g/mol so that 116.08/29.02=4 and the molecular formula is $C_4H_4O_4$.

As another example a hydrated salt of Nickel(II) chloride has the chemical formula $NiCl_2 \cdot xH_2O$ where x is the number of water molecules associated with each $NiCl_2$ unit. When 8.00g of the hydrated salt was heated, 4.36g of the anhydrous salt and 3.64g of water were produced. Determine the number of water molecules associated with $NiCl_2$ (x) in the hydrated salt. Using the format gives

Chapter 2: The Average Atomic Mass | 45

$$\text{NiCl}_2 \cdot x\text{H}_2\text{O} \rightarrow \text{NiCl}_2 + x\text{H}_2\text{O}$$

lab 8.00 g   4.36 g   3.64 g

$$\xrightarrow{\frac{1}{M_{NiCl_2}}} \downarrow \qquad \xrightarrow{\frac{1}{M_{H_2O}}} \downarrow$$

mole $n_{NiCl_2} = .0336 \xrightarrow{\frac{x}{1}} n_{H_2O} = .202$

and 0.202molH$_2$O/0.0336molNiCl$_2$=6 so that the formula of the hydrated salt is NiCl$_2 \cdot 6$H$_2$O

EXERCISES

The combustion of 3.750g of a compound containing only C and H produced 11.77g of CO$_2$ and 4.817g of H$_2$O and the molar mass of the compound was 84.16g/mol. Determine the molecular formula of this compound. (C$_6$H$_{12}$)

The combustion of 6.780g of a compound containing only C, H and O produced 15.41g of CO$_2$ and 6.309g of H$_2$O and the molar mass of the compound was 116.16g/mol. Determine the molecular formula of this compound. (C$_6$H$_{12}$O$_2$)

When 7.285g of the hydrated salt K$_2$CO$_3$·xH$_2$O was heated, 6.150g of the anhydrate salt K$_2$CO$_3$ was produced. Determine the chemical formula of the hydrated salt. (x=1.4)

46 | Stoichiometry and Beyond

A Prelude to Stoichiometry

Stoichiometry is an area of Chemistry that predicts the quantities of reactants or products in a chemical reaction given quantities of other reactants or products. The predictions are based on assuming the reaction proceeds until some compound runs out. Consider the following problem. A mass of 8.00g of methanol (CH_3OH) is combusted with 9.60g of O_2. Determine the mass of CO_2 and H_2O that will be produced. Using the previous format, one has

$$CH_3OH \;+\; O_2 \;\rightarrow\; CO_2 \;+\; H_2O$$

lab: 8.00g, 9.60g

$\downarrow \frac{1}{M_{CH_3OH}} \qquad \downarrow \frac{1}{M_{O_2}}$

mole: $n_{CH_3OH} = .25 \qquad n_{O_2} = .3$

$\downarrow \qquad\qquad \downarrow$

$n_C = .25$
$n_H = 1.0 \qquad n_O = .6$
$n_O = .25$

To proceed further, one needs to assume whether the methanol or oxygen runs out (the limiting reagent). If one assumes that the methanol runs out, then the moles of C and H on the product side are easily predicted which also determines the corresponding moles of O so that

$$CH_3OH \;+\; O_2 \;\rightarrow\; CO_2 \;+\; H_2O$$

lab: 8.00g, 9.60g

$\downarrow \frac{1}{M_{CH_3OH}} \qquad \downarrow \frac{1}{M_{O_2}}$

mole: $n_{CH_3OH} = .25 \qquad n_{O_2} = .3$

$\downarrow \qquad\qquad \downarrow$

$n_C = .25$
$n_H = 1.0 \qquad n_O = .6 \;\rightarrow\; n_C = .25 \qquad n_H = 1.0$
$n_O = .25 \qquad\qquad\qquad\qquad n_O = .5 \qquad n_O = .5$

One notices that the number of moles of O on the product side is 1.0 but that it is 0.85 on the reactant side. This assumption leads to the creation of O atoms and is therefore not correct. Therefore it is the oxygen that runs out (the limiting reagent) and some methanol is not combusted. Now the problem becomes how to determine the amount of methanol that combusted. Consider one mole of methanol. If one mole reacts, then one mole of CO_2 and two moles H_2O will be produced. Therefore in the reaction with oxygen, four moles oxygen atoms will be present in the products. Since one mole of methanol has one mole of O, three more moles of O are required or 1molCH₃OH=3molO in the reaction. Therefore, since the number of moles of O is known, one has

$$n_{CH_3OH} = n_O \left(\frac{1 mol CH_3OH}{3 mol O} \right) = .6 \left(\frac{1}{3} \right) = .2$$

so that only .2mol (6.40g) of methanol reacts leaving 0.05mol (1.60g), which did not react. Using this information gives

$$\begin{array}{cccccc}
 & CH_3OH & + & O_2 & \rightarrow & CO_2 & + & H_2O \\
\text{lab} & 6.40g & & 9.60g & & 8.80g & & 7.20g \\
 & \frac{1}{M_{CH_3OH}} \downarrow & & \frac{1}{M_{O_2}} \downarrow & & M_{CO_2} \uparrow & & M_{H_2O} \uparrow \\
\text{mole} & n_{CH_3OH} = .20 & & n_{O_2} = .3 & & n_{CO_2} = .2 & & n_{H_2O} = .4 \\
 & \downarrow & & \downarrow & & \uparrow & & \uparrow \\
 & n_C = .2 & & & & n_C = .2 & & n_H = .8 \\
 & n_H = .8 & & n_O = .6 \rightarrow & & n_O = .4 & & n_O = .4 \\
 & n_O = .2 & & & & & &
\end{array}$$

which gives the correct prediction for the masses of CO₂ and water that will be produced. As one can see, everything is consistent as far as the number of moles for each element and the conservation of mass.

If one considered the analysis that led to the number of moles of methanol that reacted somewhat arbitrary, difficult to follow and lacking insight into what to due with other chemical reactions, especially for more complicated ones, they would be absolutely correct. There clearly must be a much better way to make these kinds of predictions.

A very important way of describing chemical changes in chemistry is to use what is called a balanced chemical reaction. Consider the following description of the same chemical change discussed above.

$$2CH_3OH + 3O_2 \rightarrow 2CO_2 + 4H_2O$$

One notices that there are 2 C on each side, 8 H on each side and 8 O on each side, which is one criterion for a balanced reaction. A very important consequence of using a balanced reaction is due to the following interpretations. "When one mole of reaction occurs there are two moles of methanol. When one mole of reaction occurs there are three moles of O_2. When one mole of reaction occurs there are two moles of CO_2. When one mole of reaction occurs there are four moles of water." These statements give ten very important logical relations, which can be expressed in summary form as

$$1\,\text{molRx} = 2\,\text{molCH}_3\text{OH} = 3\,\text{molO}_2 = 2\,\text{molCO}_2 = 4\,\text{molH}_2\text{O}$$

where Rx is an abbreviation for reaction. Any pair of terms above give two conversion factors and these give rise to a total of twenty very useful conversion factors, which obviously correspond to the above reaction only. If the number of moles of methanol is known then

$$n_{Rx} = n_{CH_3OH}\left(\frac{1\,\text{molRx}}{2\,\text{molCH}_3\text{OH}}\right)$$

so that the number of moles of reaction can be determined. Similar expressions hold for other reactants or products. Similarly, if the number of moles of reaction is known then

$$n_{CO_2} = n_{Rx}\left(\frac{2\,\text{molCO}_2}{1\,\text{molRx}}\right)$$

so that the number of moles of carbon dioxide can be determined. Analogous expressions for other reactants or products are easily obtained.

The above problem can now be done with a balanced reaction. The format is the same except for the balancing coefficients and a new line. One has from the original statement of the problem that

$$2CH_3OH \;+\; 3O_2 \;\rightarrow\; 2CO_2 \;+\; 4H_2O$$

lab   8.00g   9.60g

$$\frac{1}{M_{CH_3OH}}\downarrow \qquad \frac{1}{M_{O_2}}\downarrow$$

mole $n_{CH_3OH} = .25$ $n_{O_2} = .3$

$$\frac{1\,\text{molRx}}{2\,\text{molCH}_3\text{OH}}\downarrow \qquad \frac{1\,\text{molRx}}{3O_2}\downarrow$$

LR $n_{Rx} = .125$ $n_{Rx} = .1$

and a new line has been added in the mole world, the LR line (LR is an abbreviation for limiting reagent). This line contains the number of moles of reaction based on the number of moles of the compound above it. One notices that the moles of reaction based on methanol is .125mol but based on oxygen it is .1mol. Since the reaction will proceed only until something runs out, the smallest result for n_{Rx} is the amount of reaction that can actually occur, i.e., the oxygen will run out first. Since .1mol is the actual moles of reaction that occurs, it can be put on the LR line for all compounds and one has (the units in the conversion factors between the mole and LR have been suppressed)

$$
\begin{array}{lccccc}
 & 2CH_3OH & + & 3O_2 & \to & 2CO_2 & + & 4H_2O \\
\text{lab} & 6.40\text{g} & & 9.60\text{g} & & 8.80\text{g} & & 7.20\text{g} \\
 & M_{CH_3OH} \uparrow & & \frac{1}{M_{O_2}} \downarrow & & M_{CO_2} \uparrow & & M_{H_2O} \uparrow \\
\text{mole} & n_{CH_3OH} = .2 & & n_{O_2} = .3 & & n_{CO_2} = .2 & & n_{H_2O} = .4 \\
 & \frac{2}{1} \uparrow & & \frac{1}{3} \downarrow & & \frac{2}{1} \uparrow & & \frac{4}{1} \uparrow \\
\text{LR} & n_{Rx} = .1 & & n_{Rx} = .1 & & n_{Rx} = .1 & & n_{Rx} = .1 \\
\end{array}
$$

The column under methanol predicts the mass of methanol that reacts based on the number of moles of reaction determined by the limiting reagent oxygen. One notices that, in the mole world, all conversion factors are simple ratios of integers, which emphasizes why the mole world is so useful in solving problems. Of course, the predictions from this treatment agree with the predictions that were previously obtained.

One should clearly see that the solution of this problem using a balanced reaction is much easier than the analysis based on the moles of elements. In fact, a balanced reaction completely avoids the necessity of determining the moles for each element in order to obtain a solution where no atoms are created or destroyed. The balancing coefficients now do this task. With a balanced reaction, only the moles of compounds need be considered to make predictions and this concept can be easily extended to any chemical reaction provided that it is balanced. Although there is some effort involved in balancing reactions, this effort is more than compensated for when making stoichiometric predictions.

CHAPTER 3
STOICHIOMETRY

Introduction

In the previous chapter a common format for recording answers to questions that were asked for all problems in the area was introduced. Rather than using a method based on carefully reading a problem and asking the question "Do I know how to solve it?", which leads to a dead end street if the answer is no, this method is based on questions that can be answered and whose answered are recorded in a format that is the same for all problems. The solution is obtained by moving around in the format and the ways to do this are very consistent. The final section of Chapter 2 treated the problem of predicting the amounts of products from the masses of two reactants using an unbalanced and a balanced reaction, which is in the area of Chemistry called stoichiometry. The predictions in stoichiometry are based on the assumption that the reaction will continue until something runs out which is referred to as a 100% reaction yield. It was found that making predictions with a balanced reaction was both much more systematic and easier. Here the format will be extended to treat all problems in stoichiometry so that all problems will become essentially the same. Before doing so, some general features of this problem solving technique and its correspondence to the format will be emphasized.

Passing a course in Chemistry is determined by the ability to solve problems on exams. The problems are word problems, which are "cloaked" in language and often have, as an objective, the prediction of one or more quantitative properties. Many try to achieve this ability by learning how the example problems in the text are solved and using these techniques to solve the exercises at the end of the chapter. On test, one must recognize which technique is needed to solve a problem. Confusion can arise if the test problem is stated in a way that makes this recognition difficult. An alternative way of problem solving is based on asking questions. To be useful, the questions should be applicable to all of the problems in the area of Chemistry that is being studied. Areas such as stoichiometry and chemical equilibrium have a balanced reaction, information usually consisting of measurements and well-defined objectives. These common features are turned into three questions that are asked in all problems. Instead of reading a problem carefully, one reads the problem with the intent of answering a specific question. The format provides a completely general way to record the answers to the questions so that all problems are transformed into the same format. The questions based on the common characteristics are as follows.

Question 1:   What is the Chemistry?

Question 2:   What are the measurements?

Question 3:   What are the objectives?

In the first question, the objective is to define the balanced chemical reaction and place it in the format. In order to draw generalities it is assumed that there are two reactants, A and B with stoichiometric coefficients, a and b and two products, C and D with stoichiometric coefficients c and d. This gives the following answer to the first question.

$$aA + bB \longrightarrow cC + dD$$

Lab

mole

LR

The Lab line corresponds to measurements in the lab world and the mole line corresponds to quantities in the mole world, which have units of moles. The LR line (LR is an abbreviation for limiting reagent) determines the amount of reaction that can occur. In the second question, one is looking for measurements in the problem. Care must be taken to identify what compound in the reaction the measurement refers to and the unit of the measurement. If the unit is a lab unit, such as mass, temperature, volume, etc., it is placed on the lab line under the compound it refers to. If the unit is moles, it is placed on the mole line under the compound it refers to. In the third question, one is looking for what the problem is asking for, i.e., the objectives. The objectives are placed into the format in the same way the measurements were in the form of "something equals" denoted as, for mass, m=. The order of the questions is important in that if one can not answer the first question, there is no point in going on. The following sections develop this format for use in stoichiometry. The procedures for going between the lab and mole lines depend only on the lab units and will be developed. The procedures for moving in the mole world (the mole and LR lines) are universal for all problems and will also be developed.

The next section treats the case where the lab measurements are masses. This is followed in the next section that deals with solutes where the lab measurements correspond to solute volumes and molarities. The final section considers the stoichiometry of gases where the lab measurements are partial pressures, volumes and temperatures.

Stoichiometry for Masses

The general flow of stoichiometry problems is to take measurements for a compound on the lab line and convert them into the number of moles of the compound on the mole line. Converting these into the moles of reaction on the LR line determines what will run out (i.e. the limiting reagent) which corresponds to the smallest moles of reaction. This is used to predict the moles of other compounds on the mole line. These are then converted to measurements of the compounds on the lab line in order to determine the objectives. Suppose that initially, only reactants are present and the problem gives mass measurements of m_A for A and m_B for B. The answer to the first two questions is the format.

$$aA + bB \longrightarrow cC + dD$$
$$\text{lab} \quad m_A \quad m_B$$

mole

LR

Suppose also that the objectives of the problem are to determine the masses of the products, m_C and m_D, which gives the format.

$$aA + bB \longrightarrow cC + dD$$
$$\text{Lab} \quad m_A \quad m_B \quad\quad m_C = \quad m_D =$$

mole

LR

The flow of the problem is to convert the mass of A into the moles of A (n_A) and the mass of B into the moles of B (n_B). Denoting this with arrows gives the format.

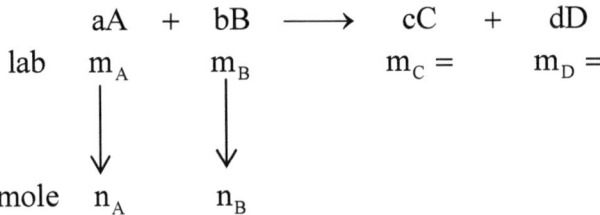

Going from the lab line to the mole line depicted by the arrow depends on the lab measurement. If the measurement is the mass of a compound X, then from the chemical formula for X one has the molar mass of X, M_X. The molar mass is defined as

$$M_X = \frac{m_X}{n_X} \tag{1}$$

where M_X is the molar mass of compound X, m_X is the mass of X and n_X is the moles of X. Then one has on rearranging Eq. (1)

$$n_X = \left(\frac{1}{M_X}\right) m_X \tag{2}$$
$$\text{mole} \quad \leftarrow \quad \text{lab}$$

where the second line indicates what lines the quantities are on in the format and what the arrow in the format is. Therefore, the arrow from the lab line to the mole line for a mass measurement is multiplication by one over the molar mass and holds for a mass measurement of any known compound. Rearranging Eq. (1) gives the arrow from the mole line to the lab line for mass as

$$m_X = (M_X)\, n_X \tag{3}$$
$$\text{lab} \quad \leftarrow \quad \text{mole}$$

or the arrow would become multiplication by the molar mass. These relations are shown below for an arbitrary compound X with stoichiometric coefficient x in a chemical reaction where the mass of X is on the lab line.

$$\begin{array}{c} xX \\ \text{lab} \quad m_X \\ \frac{1}{M_X}\downarrow \uparrow M_X \\ \text{mole} \quad n_X \end{array}$$

LR

These results are completely general for mass measurements of any known compound. One should notice that the conversions between the lab line and mole line do not, as should be obvious, involve the stoichiometric coefficient. Using this gives the format.

$$\begin{array}{ccccccc} & aA & + & bB & \longrightarrow & cC & + & dD \\ \text{lab} & m_A & & m_B & & m_C = & & m_D = \\ & \frac{1}{M_A}\downarrow & & \frac{1}{M_B}\downarrow & & & & \\ \text{mole} & n_A & & n_B & & & & \end{array}$$

LR

Going from one compound to another compound in the mole world is very simple when based on the fundamental interpretation of a balanced chemical reaction. For the reaction being considered, in one mole of reaction there are a moles of A, b moles of B, c moles of C and d moles of D. This gives very useful logical relations that can be summarized in the following way.

$$1\text{molRx} = a\text{molA} = b\text{molB} = c\text{molC} = d\text{molD} \qquad (4)$$

Any pair from Eq. (4) gives a logical relation that gives two conversion factors. If the moles of A, n_A, is known on the mole line and one wanted to determine the corresponding moles of reaction, n_{Rx}, on the LR line the logical relation is 1molRx=amolA and 1molRx/amolA would be the needed conversion factor (note that in general the moles of what you want is the term in the numerator and the moles of what you started with is the term in the denominator). Then

$$n_{Rx} = n_A \left(\frac{1 \text{molRx}}{a \text{molA}} \right) \qquad (5)$$

with analogous results for other compounds. If one knows the moles of reaction, n_{Rx}, and one wanted to know the moles of C, n_C, the logical relation is 1molRx=cmolC and cmolC/1molRx would be the needed conversion factor. Then

$$n_C = n_{Rx} \left(\frac{c \text{molC}}{1 \text{molRx}} \right) \qquad (6)$$

with analogous results for other compounds. If the moles of A is known, n_A, and one wanted the moles of B, n_B, the logical relation is amolA=bmolB and the needed conversion factor is bmolB/amolA. Then

$$n_B = n_A \left(\frac{b \text{molB}}{a \text{molA}} \right) \qquad (7)$$

with analogous results for any pair of compounds in the reaction. Then one can move from anywhere in the mole world to anywhere else in the mole world by using conversion factors with easily defined stoichiometric coefficients in the numerator and denominator (the stoichiometric coefficient of molRx is always 1). These results are summarized below.

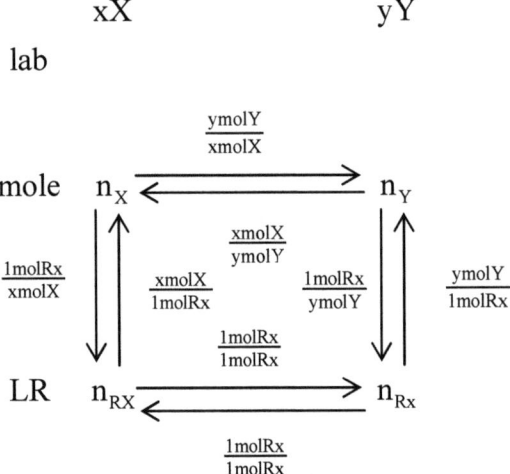

As one can see, the conversions in the mole world are very simple involving only the stoichiometric coefficients. One sees that in general the arrow is represented with integers in the numerator and denominator. The integer in the numerator is the stoichiometric coefficient of where the arrow is pointing and the integer in the denominator is the stoichiometric coefficient of where the arrow started (1 is always the stoichiometric coefficient of one mole of reaction). If X and Y are reactants and one went from the mole line to the LR line, the values of n_{Rx} on the LR line would not usually be the same since they depend on n_X and on n_Y as well as the conversion factors 1/x and 1/y. This line, as the name suggests, determines the amount of reaction that can occur. Since the reaction stops when some reactant runs out, this would correspond to the value of n_{Rx} that is the smallest (i.e., the least amount of reaction) and the corresponding compound is called the limiting reagent. This value of n_{Rx} for the limiting reagent can be placed on the LR line under each compound and used in going back to the mole line, which allows the rest of the analysis of the problem to be carried out with the limiting reagent. Since all units in the mole world are moles of something, these units, for brevity, will be suppressed in the future so that the arrows will be represented by x/y, 1/x, and so on. If you ever get confused put the units into the conversion and see if it makes sense (for example n_A(amolA/bmolB) does not make sense). Since moving along the LR line always has a conversion factor of 1/1 this will be suppressed in the future.

To determine the masses of the products you must know the moles of reaction that occurs on the LR line which gives the following format.

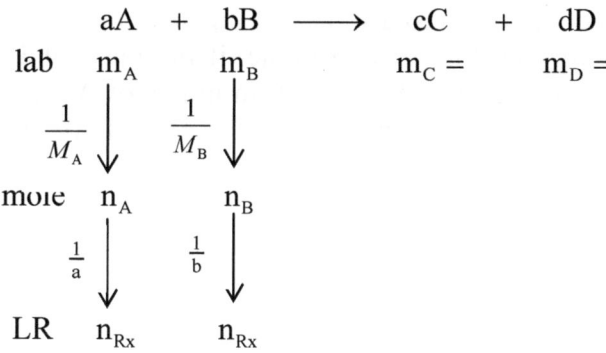

Assuming that the limiting reagent is B, i.e. the n_{Rx} value for B is smaller than for A, this can be placed on the LR line under C and D so that one can now calculate the moles of C and D and then their masses which gives the following format.

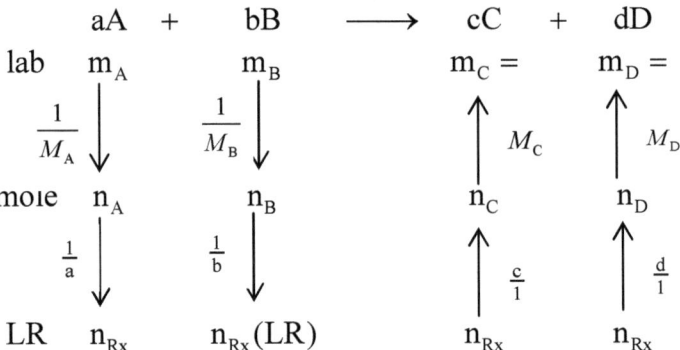

The limiting reagent problems are usually considered as the more difficult problems in stoichiometry but they are used here in the beginning of the treatment of this subject because they illustrate the importance of the LR line and how useful it is in solving problems. As a practical matter I seldom use the summary of logical relations given in Eq. (4). I simply remember that the stoichiometric coefficient of moles of reaction is one and the other stoichiometric coefficients are given in the chemical reaction part of the format. Simply use the stoichiometric coefficient of where you are going in the numerator and the stoichiometric coefficient of where you started in the denominator of the required conversion factor.

Many problems do not need to use the LR line and the conversion factors like amolA/bmolB become very useful. For example, if the mass of B (m_B) was given and the objectives were to determine how much in grams of A was needed and how much in grams of C and D were formed, then the format for the answers to the questions would become as shown below.

$$aA + bB \longrightarrow cC + dD$$

lab $m_A =$ m_B $m_C =$ $m_D =$

mole

LR

58 | Stoichiometry and Beyond

Since B is obviously the limiting reagent, you can go to the mole line for B and use n_B along with easily defined conversion factors to determine the moles of the other compounds. The masses of the compounds are then obtained. This gives the following format for the solution.

$$
\begin{array}{c}
\text{aA} \quad + \quad \text{bB} \quad \longrightarrow \quad \text{cC} \quad + \quad \text{dD} \\
\text{lab} \quad m_A = \qquad m_B \qquad\qquad m_C = \qquad m_D = \\
\Big\uparrow M_A \quad \Big\downarrow \tfrac{1}{M_B} \qquad \Big\uparrow M_C \qquad \Big\uparrow M_D \\
\text{mole} \quad n_A \longleftarrow n_B \longrightarrow n_C \quad\searrow\quad n_D \\
\tfrac{a}{b} \qquad\qquad \tfrac{d}{b}
\end{array}
$$

LR

In doing calculations, one starts with the initial quantity, in this case the mass of B, and then follows the arrows leading to the desired quantity. Each arrow along the path represents a multiplication by the conversion factor. For example, in going from the mass of B to the mass of C, there is an arrow that goes from m_B to n_B and then one that goes from n_B to n_C and finally one that goes from n_C to m_C. Thus to go from m_B to m_C there are three multiplications. The first is one over the molar mass of B, the second is c/b and the last one is the molar mass of C. One notices that the results are totally equivalent to the results using the usual technique based on a "linear analysis" which would be, for the mass of C knowing the mass of B, given by

$$m_C = m_B \left(\frac{1}{M_B}\right)\left(\frac{c}{b}\right)(M_C)$$

which is the same as following the arrows from m_B to $m_C=$ in the format above. It is felt that, in comparison to the "linear analysis" treatment, the format used here leads to a more natural way to connect measurements to objectives and easily handles multiple objectives simultaneously. One should also notice how "other types" of problems fit very naturally into this format. For example, if the moles of B were given and the mass of C produced was the objective, the starting point in the above format would be n_B on the mole line. If the mass of B needed to make a given mass of D was the objective, m_D on the lab line would be the starting point. In other words, multiple measurements and/or multiple objectives are easily handled in the format. The measurements and objectives of the problem itself define what to do.

As stated earlier, the predictions of stoichiometry are based on assuming that the reaction goes to completion, i.e., a 100% reaction yield. Most chemical reactions, however, do not go to completion. In order to treat this situation, the concept of reaction yield or percent yield is introduced in stoichiometry. The percent yield (%Y) is defined as

$$\%Y = \left(\frac{AY}{TY}\right) 100 \tag{5}$$

where AY is the actual yield of a product and TY is the theoretical yield of the same product, which is based on the result from stoichiometry. The yield of a product is usually an extensive property such as mass. For the mass of product X, one would have, for Eq. (5),

$$\%Y = \left(\frac{m_{X,act}}{m_{X,theor}}\right) 100 \tag{6}$$

where $m_{X,act}$ is the actual mass of X and $m_{X,theor}$ is the theoretical mass predicted by stoichiometry. One notices that, if the reaction goes to completion, then $m_{X,act} = m_{X,theor}$ so that the percent yield is 100% as it should be. One also notices that the actual yield can never be greater than the theoretical yield so that the percent yield is never greater than 100%.

There are several types of problems that involve percent yields. One type gives the actual mass of say C in the above reaction that is produced from a given mass of say A and the objective is to determine the percent yield. The solution is obtained by using stoichiometry to determine the theoretical mass of C and then using Eq. (6) to calculate the percent yield. Another type of problem gives the percent yield and has as an objective to determine the actual mass of C produced from a given amount of A. The solution is obtained by using stoichiometry to determine the theoretical mass of C and then rearrange Eq. (6) to

$$m_{C,act} = \left(\frac{\%Y}{100}\right) m_{C,theor}$$

in order to determine the actual mass of C. Another type of problem gives the percent yield and wants to determine the amount of a reactant, say B, that is needed to produce a given mass of a product, say D. Since the mass of D is the actual yield, Eq. (6) is rearranged to give

$$m_{D,theor} = \left(\frac{100}{\%Y}\right) m_{D,act}$$

in order to determine the theoretical yield which is needed in stoichiometry to determine the mass of B that is necessary to produce the desired amount of D. One should notice that incorporating reaction yields in problems is quite straightforward provided that one realizes that stoichiometry uses and predicts only theoretical yields.

In the case that there is more than one chemical reaction involved in the problem, one simply uses the format for each chemical reaction (Question 1), records the information in the appropriate place in the formats (Question 2) and places the objectives in the appropriate place (Question 3). Using the formats to solve the problem is the same as in the one-reaction problems.

Some examples will now be given for an actual chemical reaction and the same reaction will be used for all examples. The answer to the first question will always be the following format.

$$4NaCl_{(s)} + S_2Cl_{2(\ell)} + SF_{4(g)} \longrightarrow 4NaF_{(s)} + 3SCl_{2(\ell)}$$

lab

mole

LR

I also find it convenient to calculate all of the molar masses and store them in the calculator. The molar masses are 58.44g/mol for NaCl, 135.04g/mol for S_2Cl_2, 108.7g/mol for SF_4, 41.99g/mol for NaF and 102.97g/mol for SCl_2. These are stored in A, B, C, D and E respectively. Doing this makes moving between the lab and mole lines simple. Since the units in the mole world are always moles, it will no longer be used.

a) Determine how many moles of SCl_2 can be produced from 2.50mol of sodium chloride. The format for the solution is given below.

$$4NaCl_{(s)} + S_2Cl_{2(\ell)} + SF_{4(g)} \longrightarrow 4NaF_{(s)} + 3SCl_{2(\ell)}$$

lab

mole 2.5 $\xrightarrow{\frac{3}{4}}$ n = 1.88

LR

b) Determine how much in grams sulfur dichloride can be produced if 1.25mol of sodium chloride is reacted with 0.300mol disulfur dichloride and 0.310mol of sulfur tetrafluoride. The format for the solution is given below.

$$4NaCl_{(s)} + S_2Cl_{2(\ell)} + SF_{4(g)} \longrightarrow 4NaF_{(s)} + 3SCl_{2(\ell)}$$

lab   m = 92.7g
                                                                                        ↑ M

mole   1.25   .3   .31   .9
       $\frac{1}{4}$↓ $\frac{1}{1}$↓ $\frac{1}{1}$↓   ↑ $\frac{3}{1}$

LR   .313   .3   .31   .3

Notice that .3 was the lowest result on the LR line so that S_2Cl_2 is the limiting reagent. Thus .3 was used for SCl_2 on the LR line, which determined its moles and mass. For convenience the symbol M beside an arrow going from the mole line to the lab line indicates multiplication by the corresponding molar mass. Similarly the symbol $1/M$ beside an arrow going from the lab line to the mole line indicates a division by the corresponding molar mass.

c) Determine the mass of all compounds after 23.4g of NaCl, 20.3g of S_2Cl_2 and 10.0g 0f SF_4 completely react. The format solution is shown below.

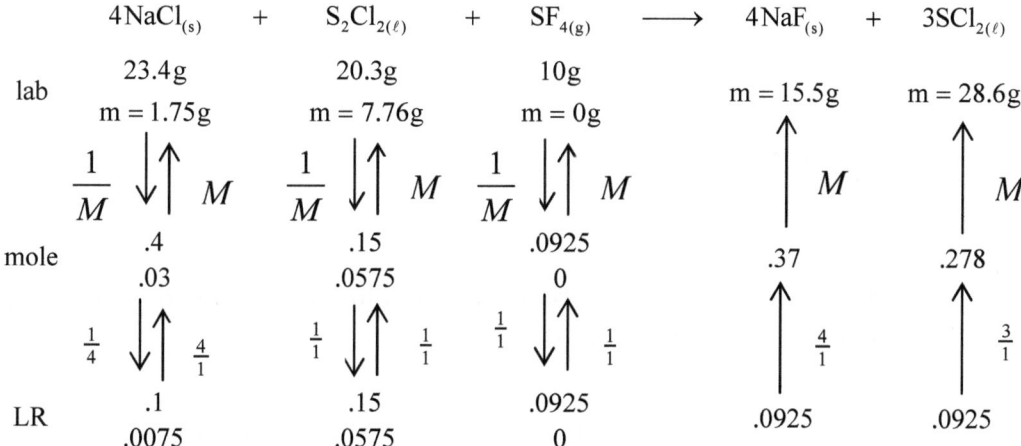

In the format above when there are two entries for a compound on a line the one above is the initial amount and the one below is the final amount. Also for the reactants the actual moles of reaction was subtracted from the n_{Rx} for the reactants, which gave the number of moles of reaction that did not occur for the reactants. This was used to determine the moles and mass of each reactant after the reaction occurred. Notice also how many objectives can be treated with a single format and that no un-needed calculations are done.

d) Determine the masses of the reactants needed to produce 57.5g of sulfur dichloride.

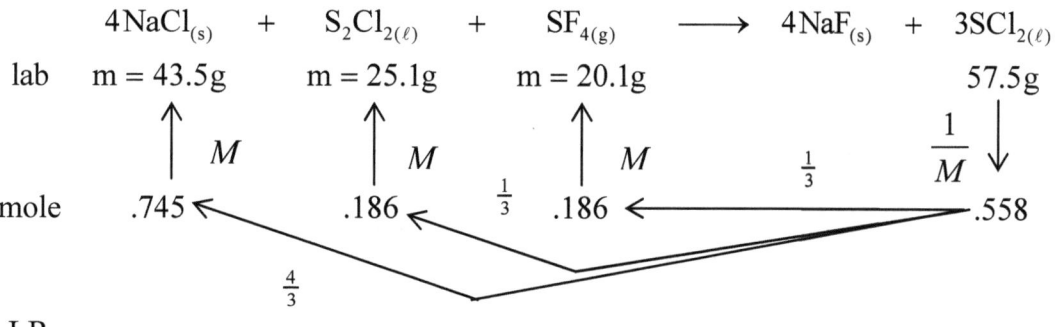

Chapter 3: Stoichiometry | 63

e) When 32.5g of NaCl reacted with excess amounts of other reactants, 33.5g of sulfur dichloride were produced. Determine the percent yield for this reaction.

From Eq. (6) and the actual mass of SCl_2 we have

$$\%Y = \left(\frac{33.5g}{m_{X,theor}}\right) 100 \ .$$

The theoretical mass is based on stoichiometry and the format solution to obtain $m_{SCl_2,theor}$ is given below.

$$4NaCl_{(s)} \ + \ S_2Cl_{2(\ell)} \ + \ SF_{4(g)} \ \longrightarrow \ 4NaF_{(s)} \ + \ 3SCl_{2(\ell)}$$

lab   32.5g   m = 42.9g

$\frac{1}{M} \downarrow$ $\frac{3}{4}$ $\uparrow M$

mole   .556 $\longrightarrow$ .417

LR

From Eq. (6) and the theoretical mass of SCl_2 we have the percent yield

$$\%Y = \left(\frac{33.5g}{42.9g}\right) 100 = 78.1 \ .$$

f) If the percent yield for this reaction is 78.1%, determine the masses of the reactants needed to make 57.5g of sulfur dichloride which was used in part d. We know that a larger mass of reactants is needed to produce an actual mass of a product than stoichiometry would predict for the actual mass. From Eq. (6), the actual mass of SCl_2 and the percent yield we have

$$m_{SCl_2,theor} = \left(\frac{100}{78.1}\right) 57.5g = 73.6g \ .$$

The format solution to obtain the masses of the reactants is shown below.

64 | Stoichiometry and Beyond

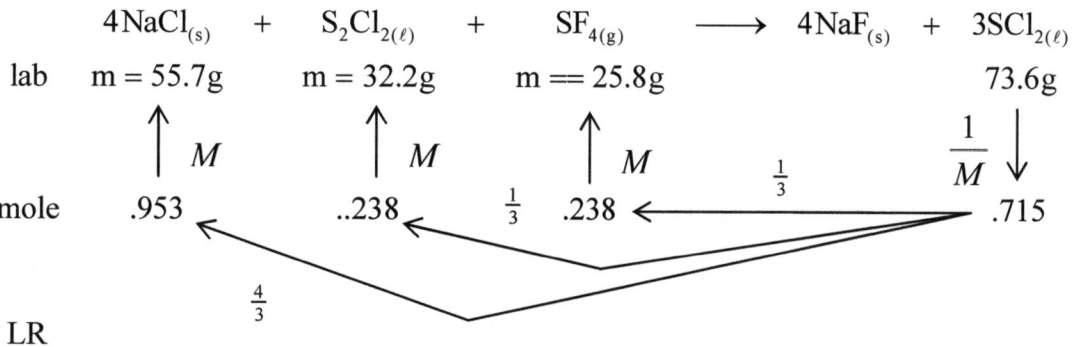

g) For a percent yield of 78.1%, determine the masses of all compounds after 20.0g of NaCl, 17.0g of S$_2$Cl$_2$ and 10.0g of SF$_4$ react. From Eq. (6) we have

$$78.1 = \left(\frac{m_{X,act}}{m_{X,theor}}\right) 100$$

and one can from the masses of the reactants determine the theoretical mass of either product, which is chosen to be NaF. Using this in the above equation would give the actual mass of NaF and this can be used in stoichiometry to give the mass of the other product and the masses after the reaction of the reactants. The format solution for the theoretical mass of NaF is shown below.

$$4NaCl_{(s)} + S_2Cl_{2(\ell)} + SF_{4(g)} \longrightarrow 4NaF_{(s)} + 3SCl_{2(\ell)}$$

lab	20g	17g	10g	m = 14.4g	
	↓ 1/M	↓ 1/M	↓ 1/M	↑ M	
mole	.342	.126	.0925	.342	
	↓ 1/4	↓ 1/1	↓ 1/1	↑ 4/1	
LR	.0856	.126	.0925	..0856	

Then

$$m_{X,act} = \left(\frac{78.1}{100}\right) 14.4g = 11.2g$$

The format solution to determine the final masses is given below.

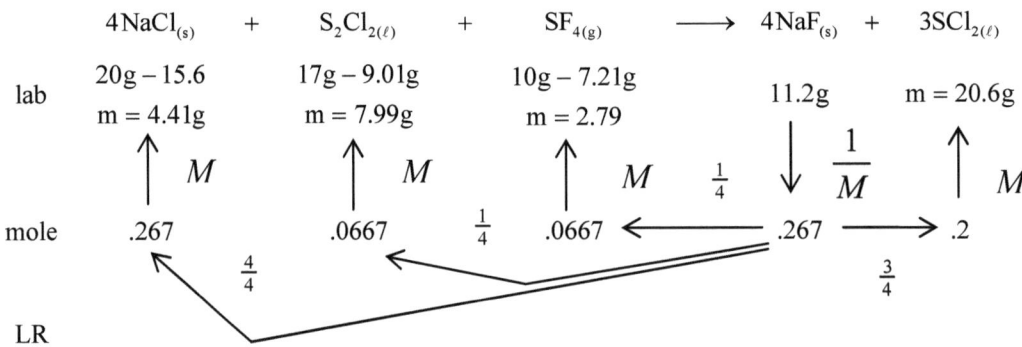

To show flexibility the final masses of reactants were obtained by subtracting the mass of the reactant that reacted from the initial mass. Note that even though NaCl is the limiting reagent its final mass is not zero. Only 100% yield reactions have a zero mass for the limiting reagent.

One should also notice that a variety of problems were done but the techniques used to solve them remained the same. The same questions were always asked. The answers always led to the same type of format. The determination of objectives always involved moving around in the format from the measurements. The rules for moving in the format were always the same. When problems had percent yields the percent yield equation which is not in the format and the format were used to develop the solution of the problem. There is no longer any need to remember how any particular problem is solved since the methods used to solve them are always the same.

EXERCISES

These exercises deal with the following reaction where A and B are the reactants and C and D are the products. The molar masses of A, B, C and D in g/mol are given above the compounds.

$$\begin{array}{cccc} 28.00 & 17.50 & 24.25 & 15.50 \\ 2A \ + \ 5B & \longrightarrow & 4C \ + & 3D \end{array}$$

Lab

Mole

LR

i) Determine the moles of C and D that are formed and the moles of A that react when 2.25mol of B react. (n_C=4.50mol)

ii) Determine the final moles of all compounds when 5.20mol of A react with 2.04mol of B. (n_C=1.63mol)

iii) Determine how many moles of A and B are needed to produce 3.75mol of D. (n_A=2.50mol)

Chapter 3: Stoichiometry | 67

iv) Determine the mass of B that reacts and the masses of C and D that are produced when 14.38g of A reacts. (m_C=24.9g)

v) Determine the masses of the reactants needed to produce 28.82g of C. (m_A=16.6g)

vi) Determine the final masses of all compounds when 17.27g of A react with 28.00g of B. (m_A=29.9g)

vii) A mixture of A and B is 41.56% A. Determine the final masses of all compounds when the mixture has a mass of 37.5g. (m_C=24.3g)

viii) Determine the masses of C and B when 2.00mol of A reacts with 4.00mol of B. (m_C=77.6g)

ix) Determine the masses of A and B needed to produce 19.65g of C if the percent yield of the reaction is 65.79%. (m_A=17.2g)

x) For the percent yield of 65.79%, determine the mass of D produced when 8.43g of A reacts with excess B. (m_D=4.61g)

xi) For a percent yield of 65.79%, determine the mass of all compounds when 72.48g of A react with 105.0g of B. (m_C=76.6g)

Stoichiometry for Solutes

A solution is defined to be a homogeneous mixture of two or more components. Typical examples of solutions are the atmosphere, which is a homogeneous mixture of gases, and seawater, which is a homogeneous mixture of water and various solvated, salts. Since solutions are homogeneous, the components are evenly mixed throughout the solution. The most common solutions in Chemistry are liquid solutions that consist of water mixed with other components. If the solution is made up of mostly one component, that component is called the solvent and the other components are called solutes. Water is one of the most common solvents in Chemistry and most of this discussion will deal with these types of solutions. A solute that is dissolved in water is called a solvated solute and in a chemical reaction written as $A_{(aq)}$ where A can be either a neutral or charged compound. If A is a charged compound, the solution is called an electrolyte solution and the charged compounds are called electrolytes. An example of an electrolyte solution is when solid sodium chloride is dissolved in water to give solvated $Na^+_{(aq)}$ and $Cl^-_{(aq)}$ ions.

Two important properties of a solution are its volume and its composition. The volume of a solution is easily determined with glassware available in the laboratory. The composition of a solution can be defined in various ways. One way could be to describe it in terms of percent masses. For a solution made up of components
A, B, C,···, the percent mass of say component A would be

$$\%A = \frac{m_A}{m_{tot}} (100)$$

where m_A is the mass of A and m_{tot} is the total mass. This may be convenient for solid solutions where the total mass is easily determined but, for liquid solutions, it could be somewhat inconvenient. If the density (d) of the liquid solution is known, then for a volume V of the solution one has

$$m_{tot} = Vd$$

so that the mass of the component A present the volume V could be determined. Notice that, if the mass of A present in the solution is known, then the moles of A is also known from

$$n_A = \frac{m_A}{M_A}$$

which gives the pathway into the mole world for solutions defined in terms of percent masses.

Another way of defining the composition of a solution that gives a more direct pathway to the mole world is called Molarity. For a solution made up of solutes A, B, C,···, the molarity of a solute, say A, is defined to be the number of moles of A present in the solution divided by the total volume of the solution in liters. Symbolically, the molarity of A, [A], is written as

$$[A] = \frac{n_A}{V} \qquad (7)$$

where n_A is the number of moles of A in the solution and V is the total volume of the solution. The unit for this concentration is called molar (M) and is defined as M=mol/L. Notice that the unit of volume in this unit is the liter. One notices that molarity is an intensive property. It is straightforward to make up solutions with a desired molarity. Consider a single solute sugar. To make a one molar aqueous sugar solution one would weigh on a scale the molar mass of sugar and then add this mass of sugar to enough water to make one liter of the solution. Notice that the molar mass of sugar was not added to one liter of water because the resulting solution may not have a total volume of one liter. Finally if this solution were stored in a bottle and one wished to distinguish it from other solutions, a label with [sugar]=1.00M would achieve this. One should notice however that the label only states how the solution was prepared, not what is actually present in the solution. This is seen more clearly by considering a one molar sodium chloride solution, which was made by adding the molar mass of NaCl to enough water to make one liter of solution. The label would be [NaCl]=1.00M but all the NaCl dissolved into $Na^+_{(aq)}$ ions and $Cl^-_{(aq)}$ so that in the solution $[Na^+_{(aq)}] = 1.00M$ and $[Cl^-_{(aq)}] = 1.00M$ which is not stated on the bottle.

A matter of practical importance occurs quite often in the laboratory. Normally there are available stock solutions of given molarity for a number of solutes. However an experiment usually requires a molarity that is not the same as the stock solution. Most stock solutions are concentrated solutions so that one needs to dilute these solutions in order to obtain the desired molarity. This procedure is referred to as dilution. The essential thing to remember in doing dilutions is that only the solvent is added to the solution so that the number of moles of the solute used to make it does not change. Suppose that one started with an initial volume (V_i) of a solution with an initial molarity of solute A of $[A]_i$ and some solvent was added to give a final volume of V_f. Since the number of moles of A, n_i, does not change, one has by rearranging Eq. (7)

$$n_{Ai} = V_i[A]_i = V_f[A]_f$$

or

$$V_i[A]_i = V_f[A]_f \qquad (8)$$

For the above situation, the initial and final concentrations are known and choosing a convenient initial volume gives

$$V_f = V_i\left(\frac{[A]_i}{[A]_f}\right)$$

which would determine what the final volume of the solution should be to obtain the desired molarity. If the final volume was known then Eq. (7) would give the needed volume of the concentrated solution. Adding the solvent to this volume until it has the final volume would give the desired solution.

The obvious advantage of using molarity to describe concentrations is seen by rearranging Eq. (7) to give

$$n_A = V[A] \qquad (8)$$

This equation says that for a solution of volume V and molarity of A of [A], the number of moles of A is determined by taking their product. V and [A] are lab world quantities and going into the mole world is very simple, even simpler than using molar masses. Similarly, if the moles of A and the volume of the solution is known, Eq. (7) determines the molarity. This is especially useful when the moles of A is determined from stoichiometry. Shown below are the mass and molarity, volume conversions between the lab and mole lines in stoichiometry.

$$\text{lab} \quad \begin{array}{c} xX \\ m_X \end{array} \qquad \begin{array}{c} xX \\ V, [X] \end{array}$$

$$M \updownarrow \quad \dfrac{1}{M} \quad V[X] \updownarrow \quad \dfrac{n_X}{V}$$

$$\text{mole} \quad n_X \qquad\qquad n_X$$

LR

The flow of problems in stoichiometry for solutions defined with molarity is the same as for problems involving masses. The only difference is in going between the lab and mole lines. If all the reactants and products in the general reaction were solutes then the answer to the first question would become as shown below.

$$aA_{(aq)} + bB_{(aq)} \longrightarrow cC_{(aq)} + dD_{(aq)}$$

lab

mole

LR

Notice that the only change is in designating the reactants and products as solutes. Suppose that a solution of A with volume V_A and of initial molarity $[A]_i$ is added to a solution of B with volume V_B and of initial molarity $[B]_i$ and the objective is to determine the final molarities of all compounds. Then the format becomes as shown below.

$$aA_{(aq)} + bB_{(aq)} \longrightarrow cC_{(aq)} + dD_{(aq)}$$

lab $V_A, [A]_i$ $V_B, [B]_i$ $V, [C]_f =$ $V, [D]_f =$
      $V, [A]_f =$ $V, [B]_f =$

mole

LR

Since the solution of A was added to the solution of B, the total volume of the solution is $V = V_A + V_B$. The flow of the problem is the same as the one for masses in that one goes to the mole line and then to the LR line to determine the amount of reaction. This is shown below.

$$aA_{(aq)} + bB_{(aq)} \longrightarrow cC_{(aq)} + dD_{(aq)}$$

lab $V_A, [A]_i$ $V_B, [B]_i$ $V, [C]_f =$ $V, [D]_f =$
      $V, [A]_f =$ $V, [B]_f =$

$V_A[A]_i \downarrow$ $V_B[B]_i \downarrow$

mole n_A n_B

         $\frac{1}{a} \downarrow$ $\frac{1}{b} \downarrow$

LR n_{Rx} n_{Rx}

Assuming that B is the limiting reagent gives $[B]_f = 0M$ and this value of n_{Rx} can be used for all compounds on the LR line which can be taken to the mole line and then to the Lab line as shown below.

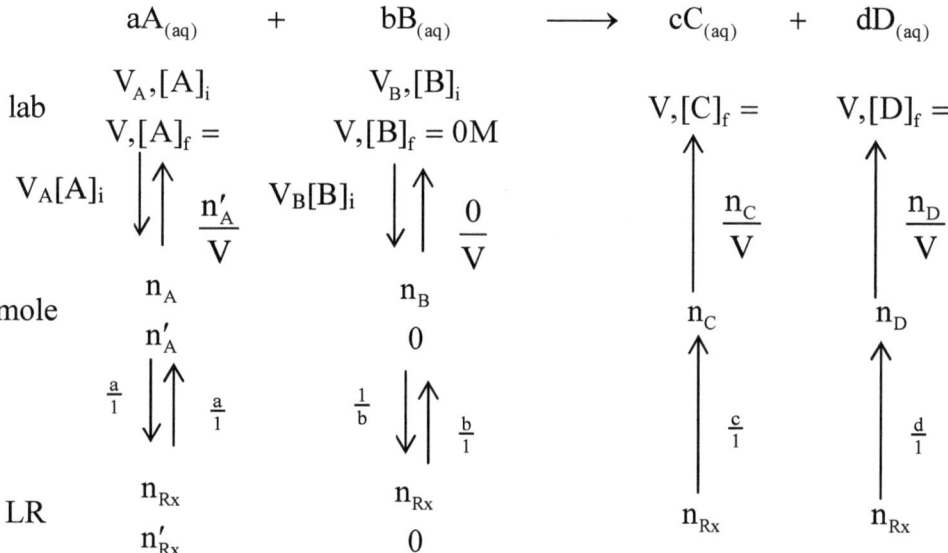

In the analysis for the final molarity of A, n'_{Rx} is the moles of reaction for A that did not occur obtained by subtracting the actual moles of reaction from the moles of reaction for A. Also n'_A is just n'_{Rx} times a/1 or the moles of A that did not react. Notice that the only difference between solids and solutes is how you go between the lab and mole lines. Movement in the mole world is completely universal.

This format can easily incorporate both solids and solutes in reactions. For example, suppose that a mass of a solid A, m_A, reacted with a solute B with initial molarity $[B]_i$ and volume V producing a solute C and a solid D and the objectives are the final masses and molarities. The answers to the three questions and the analysis to the LR line for the reactants are shown below.

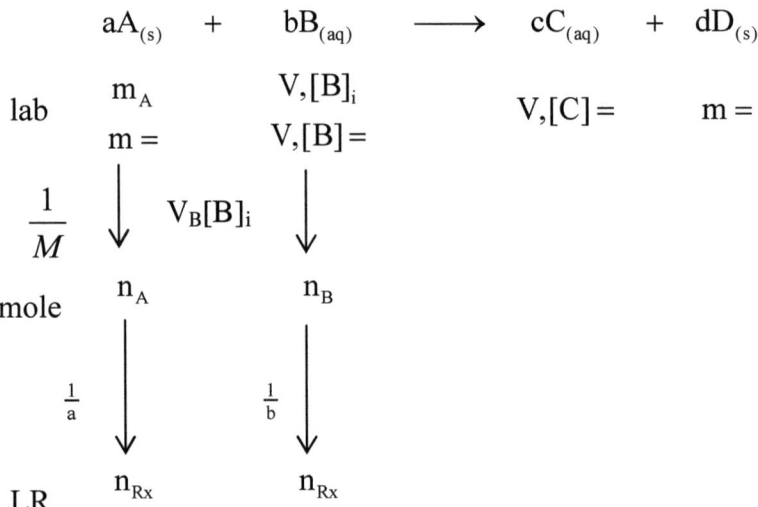

$$aA_{(s)} + bB_{(aq)} \longrightarrow cC_{(aq)} + dD_{(s)}$$

Assuming that B is the limiting reagent leads to the analysis shown below.

$$aA_{(s)} + bB_{(aq)} \longrightarrow cC_{(aq)} + dD_{(s)}$$

As is seen the format is quite flexible. Specific examples are now given dealing with the chemistry of some important types of reactions.

The percent yield for a reaction can also be defined in terms of molarities. For a solute X with mass m_X present in the volume V, one has

$$m_X = [X]VM_X$$

and when used in Eq. (6) gives after canceling

$$\%Y = \left(\frac{[X]_{act}}{[X]_{theor}}\right)100 \qquad (9)$$

where $[X]_{act}$ is the actual molarity of X and $[X]_{theor}$ is the theoretical molarity of X obtained from stoichiometry.

Thus far the answer to "What is the Chemistry?" has been obvious since a balanced reaction has been provided. Now we will consider two types of reactions called precipitation reactions and acid-base reactions both of which take place in aqueous solutions. One of the objectives is to, knowing the reactants, being able to predict the products.

In precipitation reactions, the reactants are usually solutions of soluble ionic salts forming solvated cations and anions (i.e., an electrolyte solution). Normally one pair of the cations and anions will combine to form an insoluble ionic solid (salt) and this solid is called a precipitate. For example, the salts silver nitrate, $AgNO_3$, sodium chloride, NaCl, and sodium nitrate, $NaNO_3$, readily dissolve in water. However, when a solution of silver nitrate is added to a solution of sodium chloride, a white solid (precipitate) is formed which must be the silver chloride salt, AgCl, since the only other possibility sodium nitrate is soluble. The chemical reaction for this process can be written as shown below.

$$AgNO_{3(aq)} + NaCl_{(aq)} \rightarrow AgCl_{(s)} + NaNO_{3(aq)}$$

This reaction states that, if a solution containing the soluble salt silver nitrate is added to one that contains the soluble salt sodium chloride, one will obtain a solution containing the soluble salt sodium nitrate and a precipitate silver chloride will form. A more basic description of the chemistry can be obtained by using the fact that, for example, $AgNO_{3(aq)}$ in solution is actually a solution of $Ag^+_{(aq)}$ cations and $NO^-_{3(aq)}$ anions. Using this in the above reaction gives

$$Ag^+_{(aq)} + NO^-_{3(aq)} + Na^+_{(aq)} + Cl^-_{(aq)} \rightarrow AgCl_{(s)} + Na^+_{(aq)} + NO^-_{3(aq)}$$

and upon canceling identical terms on both sides results in

$$Ag^+_{(aq)} + Cl^-_{(aq)} \rightarrow AgCl_{(s)}.$$

This reaction is called the net ionic reaction and it focuses on the actual chemical changes that take place. It also makes clear that neither the sodium ion nor nitrate ion undergo any chemical changes as is evidenced by the fact that they appear on both the reactant and product sides of the previous reaction. The ions, which do not undergo any chemical change, are called spectator ions. One should notice that it is not difficult to obtain the net ionic reaction. The starting point is the reaction defined in terms of the solutions, which corresponds to the first reaction. Then expressing the aqueous salts in terms of their cations and anions leads to the ionic form of the reaction, which corresponds to the second reaction. Finally, canceling aqueous cations and anions that appear on both sides of the reaction gives the net ionic reaction, the third reaction.

Reactions corresponding to the first and third reactions are both useful in treating precipitation reactions. The third reaction, i.e., the net ionic reaction, has the advantage of focusing on the chemical change that occurs and it is independent of what the spectator ions are. For example, calcium chloride is a soluble salt as is also calcium nitrate. If a solution of silver nitrate is added to one of calcium chloride, the reaction would become

$$2\,AgNO_{3(aq)} + CaCl_{2(aq)} \rightarrow 2\,AgCl_{(s)} + Ca(NO_3)_{2(aq)}$$

and upon writing in terms of solvated cations and anions gives

$$2\,Ag^+_{(aq)} + 2\,NO^-_{3(aq)} + Ca^{2+}_{(aq)} + 2\,Cl^-_{(aq)} \rightarrow 2\,AgCl_{(s)} + Ca^{2+}_{(aq)} + 2\,NO^-_{3(aq)}$$

which after canceling spectator ions and dividing by two to give the lowest set of integers gives

$$Ag^+_{(aq)} + Cl^-_{(aq)} \rightarrow AgCl_{(s)}$$

which is the same net ionic reaction.

The first reaction is usually more useful for stoichiometry problems. Most of these problems give measurements that correspond to molarities found on bottles, which are prescriptions for how the solution is made. If the solution is silver nitrate, the measurement would most likely be [AgNO$_3$] and similarly for other solutions. The first reaction has the advantage of being written in terms of these solutions so that the measurements in the problem can be used directly in this reaction. Stoichiometry can then be used to determine the final concentration of all solvated salts and the mass of the precipitate. The concentration of the solvated cations and anions can then be found from conversion factors based on chemical formulas. This will be shown in an example.

In order to predict whether a precipitate will form and if so what it is, it is helpful to have a set of solubility rules that will give a rough guide in making these predictions. The following rules focus on either the cation or anion that could be present in a salt. If the ion is present in the salt, it is considered to be soluble unless it is combined with one of the ions listed in the exceptions.

ions	exceptions
NH_4^+	
Group IA ions	
NO_3^-	
Cl^-, Br^-, I^-	Ag^+, Hg_2^{2+}, Pb^{2+}
SO_4^{2-}	$Ba^{2+}, Hg_2^{2+}, Pb^{2+}, Sr^{2+}$

All other salts are to be considered insoluble. Note that with these rules, the silver chloride precipitate could be predicted. Notice also that these rules provide a shortcut to obtaining the net ionic reaction. For example, if one knows that the precipitate formed is AgCl, then the net ionic reaction must be

$$Ag^+_{(aq)} + Cl^-_{(aq)} \rightarrow AgCl_{(s)}$$.

As another example, if the solutions were of sodium carbonate and iron (II) chloride, these rules would predict that the precipitate would be iron (II) carbonate with the net ionic reaction of

$$Fe^{2+}_{(aq)} + CO_3^{2-}{}_{(aq)} \rightarrow FeCO_{3(s)}$$.

To gain experience with these rules, the following table should be completed. The soluble salts going across and down represent solutions of these salts. When one going across is combined with one going down, the entry underneath the one going across and the one going down should indicate either no reaction (NR) or the precipitate(s) that are formed. Note that only the upper part above the diagonal needs to be completed. It is also instructive to write down the net ionic reaction for any precipitate that is formed. Doing this will accomplish two things. First it will put the solubility rules where they belong, which is in your mind. Second it will make you much more comfortable with describing chemical reactions in this area of Chemistry.

	NaCl	NH$_4$OH	Ba(NO$_3$)$_2$	Ag$_2$SO$_4$	Pb(NO$_3$)$_2$	Na$_2$CO$_3$	(NH$_4$)$_2$S	CrCl$_3$
NaCl								
NH$_4$OH								
Ba(NO$_3$)$_2$								
Ag$_2$SO$_4$								
Pb(NO$_3$)$_2$								
Na$_2$CO$_3$								
(NH$_4$)$_2$S								
CrCl$_3$								

The stoichiometric treatment of precipitation reactions is quite straightforward. Suppose that 40.mL of a 0.40M iron (III) nitrate solution is added to 60.mL of a 0.50M sodium carbonate solution and the mass of the precipitate if any and final concentrations of all ions are to be determined. According to the solubility rules, one predicts that the precipitate iron (III) carbonate will form and the answer to the three questions and evaluation of the limiting reagent is shown in the format below.

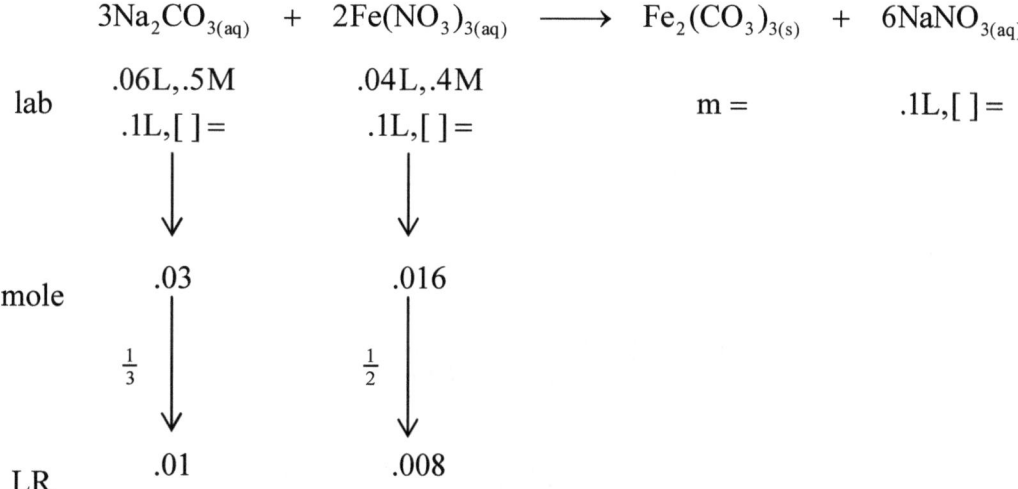

Defining the conversion factors for the arrows between the lab and mole lines is left as an exercise. Since iron (III) nitrate is the limiting reagent .008mol of reaction can be used for all compounds on the mole line and one obtains the result shown below.

$$3Na_2CO_{3(aq)} + 2Fe(NO_3)_{3(aq)} \longrightarrow Fe_2(CO_3)_{3(s)} + 6NaNO_{3(aq)}$$

lab
- .06L, .5M / .1L, [ ] = .06M
- .04L, .4M / .1L, [ ] = 0M
- m = 2.3g
- .1L, [ ] = .48M

mole
- .03 / .006
- .016 / 0
- .008
- .048

Conversion factors: $\frac{1}{3}$, $\frac{3}{1}$; $\frac{1}{2}$, $\frac{2}{1}$; $\frac{1}{1}$; $\frac{6}{1}$

LR
- .01 / .002
- .008 / 0
- .008
- .008

Notice that the final molarity of sodium nitrate is based on the total volume, which is 0.10L. Since the final molarities of the compounds are now known, the molarities of the anions and cations can be obtained from conversion factors based on the chemical formula. For sodium nitrate one has that

Chapter 3: Stoichiometry | 81

$$[NO_3^-] = \frac{n_{NO_3^-}}{V} = \frac{n_{NaNO_3}}{V}\left(\frac{1 mol NO_3^-}{1 mol NaNO_3}\right) = 0.48 M$$

and similarly

$$[Na^+] = 0.48 M \quad .$$

For the sodium carbonate one has

$$[CO_3^{2-}] = (0.06 M)\left(\frac{1 mol CO_3^{2-}}{1 mol Na_2CO_3}\right) = 0.06 M$$

and

$$[Na^+] = (0.06 M)\left(\frac{2 mol Na^+}{1 mol Na_2CO_3}\right) = 0.12 M \quad .$$

Since both the sodium carbonate and sodium nitrate contribute to the sodium ion concentration, the sum of the concentrations would be the total concentration so that

$$[Na^+] = 0.48 M + 0.12 M = 0.60 M$$

which gives the final concentrations of all ions present in solution along with the mass of 2.3g for the precipitate. Another way to determine spectator ion concentrations is to notice that neither Na^+ or NO_3^- are in the precipitate so that they must be spectator ions. The moles of these ions divided by the final volume then give the final molarities. . Therefore, since the moles of spectator ions do not change in the reaction, one has for the final concentrations

$$[Na^+] = \frac{2(.6M)(.05L)}{0.10L} = 0.60 M$$

and

$$[NO_3^-] = \frac{3(.4M)(.04L)}{0.10L} = 0.48 M \quad .$$

This problem could also be solved with the net ionic reaction. The net ionic reaction involves only the formation of the precipitate and the final molarities for spectator ions, which are easily identified, are determined as shown above. Using the molarity of carbonate ions and iron (III) ions in the format for the net ionic reaction gives the results shown below.

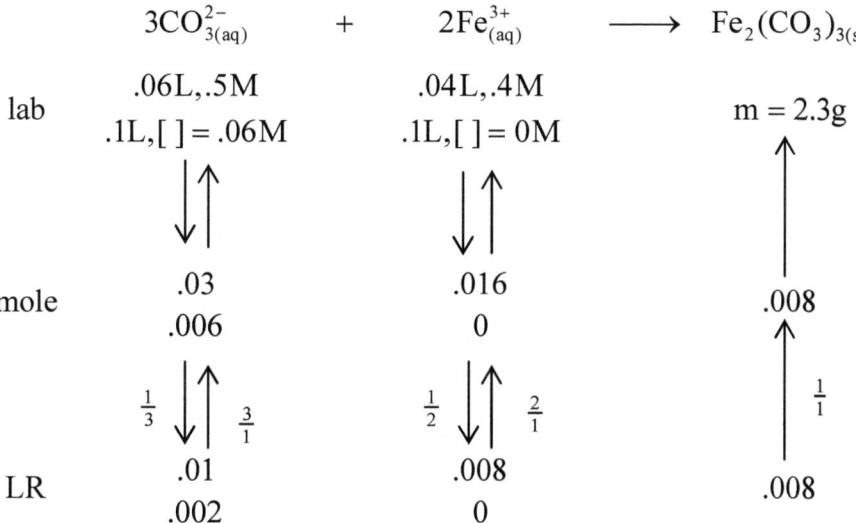

In my opinion the problem can be analyzed with either reaction but the net ionic reaction leads to a more direct solution.

The second type of reaction to be considered in this section is acid base reactions. Solutions of acids and bases are very important in the field of chemistry. Most of these solutions have water as the solvent and only these types of solutions will be considered here.

As a starting point for developing the chemistry of acids it is convenient to begin with distilled or de-ionized water. In distilled water, water molecules react to a small extent to give hydronium and hydroxide ions. Since water is neutral, the concentrations of these ions are equal. The chemical reaction that produces these ions is

$$2H_2O_{(\ell)} \rightleftharpoons H_3O^+_{(aq)} + OH^-_{(aq)}$$

where the double arrow indicates that both the forward and reverse reactions occur and do so at such a rate that all concentrations remain constant, which is referred to as a state of dynamic equilibrium. Dynamic emphasizes that both reactions are still occurring but they cancel each other out. At equilibrium, the concentrations of the hydronium and hydroxide ions are constant and one has that

$$[H_3O^+]_{eq}[OH^-]_{eq} = K_w$$

where K_w is the equilibrium constant for this reaction which is 10^{-14} at 25 °C and the subscript eq means the concentration at equilibrium. Since the solution is neutral, $[H_3O^+]_{eq}=[OH^-]_{eq}$ and using the value of K_W gives $[H_3O^+]_{eq}=[OH^-]_{eq}=10^{-7}M$ for pure water. The most interesting property of the equation above is that it applies to all aqueous solutions, not just pure water. For example, if one added something to pure water that increased the hydronium ion concentration, the hydroxide ion concentration would decrease until the equation above was again satisfied. All reactions in dynamic equilibrium have an equilibrium constant and this constant is used in predicting equilibrium concentrations. If the equilibrium constant is large, it means the forward reaction dominates or that the equilibrium mixture contains mostly products. If it is small, however, it means that the equilibrium mixture will contain mostly reactants. Since $K_w=10^{-14}$ which is small, it is no surprise that the ion concentrations are very small. Finally, if the above reaction is turned around becoming

$$H_3O^+_{(aq)} + OH^-_{(aq)} \rightleftharpoons 2H_2O_{(\ell)}$$

the new equilibrium constant is $1/K_w$ which is 10^{14} or very large. This means that this reaction goes nearly to completion to the product side.

In defining acidic and basic solutions, it is convenient to use as a reference pure water where $[H_3O^+]_{eq}=[OH^-]_{eq}=10^{-7}M$. Then an acidic solution is a solution where the $[H_3O^+]_{eq}([OH^-]_{eq})$ is greater (less) than $10^{-7}M$. Similarly, a basic solution is a solution where $[H_3O^+]_{eq}([OH^-]_{eq})$ is less (greater) than $10^{-7}M$.

Defining acids and bases is somewhat more complicated. They could be defined by what effect they had on the concentrations of hydronium and hydroxide ions in pure water. This is the essence of the Arrhenius definition in that an acid is defined as a compound that, when added to distilled water, increases the hydronium ion concentration. Similarly a base is defined as a compound that when added to distilled water increases the hydroxide ion concentration. However these definitions do not reveal any common characteristics from a molecular point of view. In 1923, a theory was proposed by Bronsted and Lowry that is very useful to understand the chemistry of aqueous solutions of acids and bases. In their theory, an acid is a proton (H^+) donor and a base is a proton acceptor. Therefore, according to this theory, the transfer of a proton from an acid to a base characterizes an acid base reaction. For example, a solution of hydrochloric acid is given by the reaction

$$\textbf{\textit{H}}Cl_{(aq)} \;+\; H_2O_{(\ell)} \;\rightleftharpoons\; Cl^-_{(aq)} \;+\; \textbf{\textit{H}}OH^+_{2(aq)}$$

forward role   acid   base

reverse role   base   acid

where the transferred proton is designated as **H** and the role of the reactant molecules in both the forward and reverse directions is indicated below. One notices that the HCl on the reactant side is an acid but when it loses a proton, on the product side Cl⁻ is a base. The pair, HCl and Cl⁻ is called a conjugate acid base pair. Similarly, H_2O (base) and H_3O^+(acid), i.e. $\textbf{\textit{H}}OH_2^+$ are a conjugate acid base pair. In summary, an acid (base) on the reactant side produces its conjugate base (acid) on the product side via a proton transfer. The same concepts apply to basic solutions such as ammonia, which has the chemical reaction

$$NH_{3(aq)} \;+\; \textbf{\textit{H}}OH_{(\ell)} \;\rightleftharpoons\; \textbf{\textit{H}}NH^+_{3(aq)} \;+\; OH^-_{(aq)}$$

forward role   base   acid

reverse role   acid   base

where again the conjugate acid base pairs are easily recognized. One should note that water acts as an acid when it reacts with a base and it acts like a base when it reacts with an acid. This theory is quite helpful in interpreting the beginning water reaction when written as

$$\textbf{\textit{H}}OH_{(\ell)} \;+\; H_2O_{(\ell)} \;\rightleftharpoons\; \textbf{\textit{H}}OH^+_{2(aq)} \;+\; OH^-_{(aq)}$$

in that one of the water molecules plays the role of an acid while the other plays the role of a base.

    It is straightforward to use the Bronsted and Lowry Theory to obtain a generalization of the chemistry of solutions of acids and bases. To do this a monoprotic acid is denoted a HA where A is the rest of the chemical formula and corresponds to the uncharged conjugate base. Monoprotic means that only one H^+ can be transferred. For nitric acid, A would be NO_3 and for acetic acid, CH_3COOH, A would be CH_3COO, etc.. Then the chemical reaction for any monoprotic acid would be

$$HA_{(aq)} + H_2O_{(\ell)} \rightleftharpoons H_3O^+_{(aq)} + A^-_{(aq)}$$

so that all monoprotic acid chemistry is the same with the exception of what A is. The products in a monoprotic reaction are always the hydronium ion and the conjugate base. For a diprotic acid of the form H_2A the chemistry for the loss of the first proton would be

$$H_2A_{(aq)} + H_2O_{(\ell)} \rightleftharpoons H_3O^+_{(aq)} + HA^-_{(aq)}$$

and for the loss of the second proton it would be

$$HA^-_{(aq)} + H_2O_{(\ell)} \rightleftharpoons H_3O^+_{(aq)} + A^{2-}_{(aq)}$$

and adding these reactions finally gives

$$H_2A_{(aq)} + 2H_2O_{(\ell)} \rightleftharpoons 2H_3O^+_{(aq)} + A^{2-}_{(aq)}$$

as the reaction for the loss of two protons. For a base of the form B where for ammonia B would be NH_3 one has for the chemistry

$$B_{(aq)} + H_2O_{(\ell)} \rightleftharpoons OH^-_{(aq)} + HB^+_{(aq)}$$

which is the chemistry of any compound that accepts a proton. One should note that some basic solutions are produced by adding a soluble hydroxide salt such as NaOH to water. In these cases the hydroxide ions are produced by the salt dissolving, not by the above reaction.

Acids and bases are classified as being either strong or weak. A strong acid is an acid that completely loses its proton. For a weak acid all acid molecules do not lose a proton. Most acids are weak. Common strong monoprotic acids are hydrochloric acid, HCl, hydrobromic acid, HBr, hydroiodic acid, HI, nitric acid, HNO_3 and perchloric acid $HClO_4$. The most common strong diprotic acid is sulfuric acid, H_2SO_4, where strong means that it completely loses its first proton. Acetic acid is a typical weak acid. A strong base completely produces the hydroxide ion. Soluble hydroxide salts such as NaOH, LiOH and $Ca(OH)_2$ are strong bases. A weak base does not completely produce the hydroxide ion and most bases are weak. Ammonia is a common example of a weak base.

When an acidic solution is added to a basic solution, the chemical reaction that takes place is called neutralization. Consider the reaction for the monoprotic acid HA

$$HA_{(aq)} + H_2O_{(\ell)} \rightleftharpoons H_3O^+_{(aq)} + A^-_{(aq)}$$

and add to the reaction

$$H_3O^+_{(aq)} + OH^-_{(aq)} \rightleftharpoons 2H_2O_{(\ell)}$$

which gives

$$HA_{(aq)} + OH^-_{(aq)} \rightleftharpoons H_2O_{(\ell)} + A^-_{(aq)} \ .$$

This reaction would describe the addition of a solution of a strong base to an acid solution and one sees that the products are water and the solvated conjugate base of the acid. If adding sodium hydroxide to water made the solution of the strong base, this could also be written as

$$HA_{(aq)} + NaOH_{(aq)} \rightleftharpoons H_2O_{(\ell)} + NaA_{(aq)}$$

where water and the soluble salt NaA are the products. This reaction has the advantage of using chemicals found on laboratory bottles. If all salts are soluble, the net ionic reaction is found to be

$$H_3O^+_{(aq)} + OH^-_{(aq)} \rightleftharpoons 2H_2O_{(\ell)}$$

which is the net ionic reaction for neutralization. Similarly, if the reaction for the base B

$$B_{(aq)} + H_2O_{(\ell)} \rightleftharpoons OH^-_{(aq)} + HB^+_{(aq)}$$

is added to the reaction

$$H_3O^+_{(aq)} + OH^-_{(aq)} \rightleftharpoons 2H_2O_{(\ell)}$$

one obtains

$$B_{(aq)} + H_3O^+_{(aq)} \rightleftharpoons H_2O_{(\ell)} + HB^+_{(aq)}$$

which corresponds to the neutralization reaction of a base with a strong acid. In terms of an actual strong acid, say HCl, the reaction would be

$$B_{(aq)} + HCl_{(aq)} \rightleftharpoons HB^+_{(aq)} + Cl^-_{(aq)}$$

or

$$B_{(aq)} + HCl_{(aq)} \rightleftharpoons HBCl_{(aq)}$$

which again is written in terms of laboratory solutions.

In the lab one has available stock solutions of acids and bases and the concentrations given on the bottle indicate how the solution was made, not what is present in the solution. For example, if the label on a bottle of nitric acid says [HNO_3]=1.00M, this does not mean that one mole of HNO_3 is present in one liter of solution. Rather it means that one mole of HNO_3 was added to enough water to make one liter of solution. Since nitric acid is a strong acid there actually are no moles of HNO_3 present but rather one mole of both hydronium ions and nitrate ions present in one liter of solution. Acetic acid is a weak acid and again the label [CH_3COOH]=1.00M indicate how the solution was prepared, not what is actually present.

Suppose now that one has an acidic solution of HA but that [HA], the prescription for how the solution was made was not known, i.e. the concentration on the bottle was not given. The experimental procedure for determining this concentration is called a titration. A titration always involves two solutions, one which is acidic and another which is basic. For the case considered here, one would measure a volume of the acid, V_{HA}, and then add an indicator that changes color at the equivalence point of the titration. The basic solution with known molarity would then be added to the acidic solution until the resulting solution changes color. This point called the equivalence point (or end point) is defined as when the number of moles of the base (OH⁻) is equal to the maximum number of moles of hydronium ions (H⁺) that the acid can lose. If the strong base is sodium hydroxide, for a monoprotic acid this would be

$$n_{NaOH} = n_{HA}$$

or for a diprotic acid

$$n_{NaOH} = 2n_{H_2A}$$

where the number of moles of the acid is the number that would be used to prepare the solution with the already measured volume, V_{HA}, and the number of moles of NaOH is obtained by multiplying the volume of the base used in the titration by the known molarity of the base. Note that the factor of two for the diprotic acid is due to the fact that each acid molecule can lose two protons. It will be shown that the equivalence point is easily obtained from the format of the stoichiometric treatment of titrations.

As an example, consider a 50.mL solution of acetic acid (CH_3COOH or HAc) with unknown molarity and 65mL of a 0.75M solution of sodium hydroxide was needed to reach the equivalence point. The objective is to determine the molarity of the initial acid solution. The format in terms of the actual solutions is shown below.

$$HAc_{(aq)} + NaOH_{(aq)} \longrightarrow H_2O_{(aq)} + NaAc_{(aq)}$$

lab .05L, [HAc] = .98M   .065L, .75M

mole   .0488 ⟵ $\frac{1}{1}$ .0488

LR

One notices that the arrow from the moles of NaOH to the moles of HAc, 1/1, is the equivalence point definition for monoprotic acids.

As a final example, suppose that a 75mL solution of sulfuric acid of unknown concentration required 95mL of a 0.45M sodium hydroxide solution to reach the equivalence point and one wanted to determine the concentration of the acid. The set up and solution for this problem is shown below.

$$H_2SO_{4(aq)} + 2NaOH_{(aq)} \longrightarrow 2H_2O_{(aq)} + Na_2SO_{4(aq)}$$

lab .075L, [HAc] = .29M   .095L, .45M

mole .0214 $\xleftarrow{\frac{1}{2}}$ .0428

LR

One notices that the factor of 1/2 in going from the moles of base to the moles of acid on the mole line is the definition of the equivalence point for a diprotic acid.

The titration of an unknown base with a known strong acid is done most straightforwardly by using the reaction

$$B_{(aq)} + HA_{(aq)} \longrightarrow HBA_{(aq)}$$

where HA is a strong acid. Using this form for the chemistry, which involves lab solutions, allows one to use the measurements directly in the format to determine the objective.

EXERCISES

1) These exercises deal with the following reaction where A and B are the reactants and C and D are the products. The molarities of A and B and molar mass of C in g/mol are given above the compounds.

$$\begin{array}{cccc} 1.35\text{M} & 1.85\text{M} & 24.25 & \\ 2A_{(aq)} & + \ 5B_{(aq)} & \longrightarrow \ 4C_{(s)} & + \ 3D_{(aq)} \end{array}$$

Lab

Mole

LR

i) Determine the moles of C and D that are formed and the moles of B that react when 2.25mol of A react. (n_C=4.50mol)

ii) Determine the final moles of all compounds when 2.04mol of A react with 5.20mol of B. (n_C=2.04mol)

iii) Determine how many moles of A and B are needed to produce 3.75mol of C. (n_A=1.88mol)

iv) Determine the volume of A, the mass of C and molarity of D that is produced when 37.5mL of B reacts. ([D]=0.714M)

v) Determine the volumes of the reactants needed to produce 28.82g of C. (V_A=0.440L)

vi) Determine the final molarities of A, B and D and the mass of C when 54.0mL of B reacts with 97.5mL of A. ([D]=0.717M)

vii) Determine the mass of C and the molarity of D when 1.80mol of A reacts with 4.00mol of B and the volume of the solution is 93.4mL. ([D]=25.7M)

viii) When 42.7mL of A reacted with 132mL of B the actual molarity of D was 0.365M. Determine the percent yield for this reaction. (73.7%)

ix) A precipitate forms when 35.4mL of 0.856M solution of $CrBr_3$ is added to 22.5mL of a 0.472M solution of NaOH. Determine the mass of the precipitate and the final concentrations of all ions. (m=8.82g)

x) A precipitate forms when 14.2mL of 0.265M solution of $Fe_2(SO_4)_3$ is added to 28.7mL of a 0.333M solution of BaI_2. Determine the mass of the precipitate and the final concentrations of all ions. (m=2.23g)

xi) Determine the volume of a 4.78M NaOH solution needed to neutralize 46.8mL of a 9.67M solution of HNO_3. (94.7mL)

xii) Determine the volume of a 0.785M NaOH solution needed to neutralize 29.6mL of a 0.467M solution of H_2SO_4. (35.2mL)

Stoichiometry for Gases

Four properties are needed to specify the state of a gas. These properties are pressure (P), volume (V), the absolute temperature (T) and the moles (n). Although one deals with real gases such as oxygen (O_2) it turns out that in most situations of interest real gases act like ideal gases and we will therefore consider all gases to be ideal gases.

In order to obtain an equation of state for an ideal gas it is convenient in studying a gas to find out how one if its properties (chosen to be V) depend on the other properties. When investigating this dependence the other two properties remain constant, which hopefully results in a simple law. One therefore expects to obtain a law for V vs. P, a law for V vs. T and another law for V vs. n. The first experimental gas law was discovered in 1662 by Robert Boyle and is known as Boyle's Law. He found that when the temperature and moles of the gas were held constant, the product of the volume times the pressure was constant or

$$PV = K_B$$

where K_B is a constant for constant temperature and amount with units of Latm. A gas that obeys this equation for all pressures is called an ideal gas. Obviously an ideal gas does not actually exist but most gases under reasonable conditions behave as if they were ideal gases. Since this holds for all states of a gas, a state with initial values of volume and pressure of V_i and P_i and another state with final values V_f and P_f are related to each other under these conditions by

$$V_i P_i = V_f P_f \qquad (10)$$

The second experimental gas law was discovered by Jacques Charles in 1787 and is known as Charles's Law. In his studies, the moles of gas and the pressure of the gas were held constant and he found that the volume of the gas was proportional to the absolute temperature or

$$V = K_C T$$

where K_C is a constant under the conditions of constant P and n and T in degrees Kelvin is defined as

$$T(K) = \left(\frac{1K}{1^\circ C}\right) t(^\circ C) + 273.15K$$

where t is the Centigrade temperature. If the gas goes from an initial state to a final state under constant pressure and amount of gas, Charles' Law takes the form

$$\frac{V_i}{T_i} = \frac{V_f}{T_f} \quad . \tag{11}$$

The final gas law was produced by Gay-Lussac and Amadeo Avogadro and is known as Avogadro's Law. This law states that, if T and P of the gas are held constant, the volume that the gas occupies is proportional to the number of moles or

$$V = K_A n \quad .$$

Under the conditions of constant temperature and pressure, the initial values of V and n are related to the final values by

$$\frac{V_i}{n_i} = \frac{V_f}{n_f} \quad . \tag{12}$$

Although these laws provide information about the gas, one would like to obtain a more general expression for a gas changing its state and the equation of state for the gas. This can be accomplished by exploiting what properties of matter are. For something to be a property of matter it must be applicable to all matter and it must not depend on how the property achieved its value (there must be no history of the matter in a value of a property). Consider an initial state of an ideal gas with values of its properties of V_i, P_i, T_i and n_i and a final state of the gas with the properties V_f, P_f, T_f and n_f. The objective is to change the state of the gas from its initial state to its final state and find a mathematical expression for this. Since V, P, T and n are properties, this change can be done in any way such that the final state is achieved. Consider performing the following three steps. First hold $T=T_i$ and $n=n_i$ constant and change the pressure of the gas to its final value P_f. These conditions correspond to Boyle's Law so that

$$V_i P_i = V_1 P_f$$

or

$$V_1 = \frac{V_i P_i}{P_f}$$

where V_1 is the final volume of the gas in the first step. In the second step, hold $P=P_f$ and $n=n_i$ constant and change T to T_f. These are the conditions of Charles' Law so that

$$\frac{V_1}{T_i} = \frac{V_2}{T_f}$$

or

$$V_2 = \left(\frac{T_f}{T_i}\right) V_1$$

where V_2 is the final volume in step two. Using V_1 from above gives

$$V_2 = \left(\frac{T_f}{T_i}\right)\left(\frac{P_i}{P_f}\right) V_i \; .$$

In the final step, $P=P_f$ and $T=T_f$ are held constant and n is changed to n_f which corresponds to the conditions of Avogadro's Law. Since this state has the values of P_f, T_f and n_f, it must have the unique volume of V_f so that

$$\frac{V_2}{n_i} = \frac{V_f}{n_f}$$

or

$$V_f = \left(\frac{n_f}{n_i}\right) V_2 \; .$$

Using the V_2 from above gives

$$V_f = \left(\frac{n_f}{n_i}\right)\left(\frac{T_f}{T_i}\right)\left(\frac{P_i}{P_f}\right) V_i \; .$$

Rearranging the above equation so that all initial state properties are on one side and all final state properties are on the other gives

$$\frac{V_i P_i}{n_i T_i} = \frac{V_f P_f}{n_f T_f} \; .$$

This equation is the most general form for changes in the state of an ideal gas. It contains Bolye's Law, Charles', Avogadro's Law and others such what the relation is between P and n when T and V are constant. Since the left hand side only involves properties of an arbitrary initial state and the right hand side only involves properties of an arbitrary final state, this can only be satisfied if each side of this equation is equal to a constant or

$$\frac{V_i P_i}{n_i T_i} = \frac{V_f P_f}{n_f T_f} = R$$

where R is a universal constant called the gas constant. Letting V, P, n and T represent an arbitrary state of the gas gives

$$PV = nRT \qquad (13)$$

which is called the ideal gas equation of state. One notices that, if three properties are known, this equation predicts the value of the fourth property, which is what the job of an equation of state is. Rearranging this equation to give

$$n = \frac{PV}{RT}$$

indicates the possibility of going from lab measurements to moles n in stoichiometry. The problem is that reactions have individual gases for reactants or products but in a lab they well could be components of a mixture of gases. Consider a mixture of gases A and B consisting of n_A moles of A and n_B moles of B in a container with volume V and temperature T. Since the total number of moles of gas is $n_{tot} = n_A + n_B$, one has

$$P_{tot} = \frac{n_{tot} RT}{V} = \frac{n_A RT}{V} + \frac{n_B RT}{V} \quad .$$

However

$$P_A = \frac{n_A RT}{V}$$

and

$$P_B = \frac{n_B RT}{V}$$

so that

$$P_{tot} = P_A + P_B$$

Chapter 3: Stoichiometry | 97

where P_A is the partial pressure of A and P_B is the partial pressure of B. One sees that the total pressure of a mixture of two gases is a sum of two pressures. The first pressure term is a pressure due to only gas A and the second is a pressure due to only gas B. In general, it is straightforward to show that if a gas mixture contains the gases $A, B, C, \cdots$, the total pressure is

$$P_{tot} = P_A + P_B + P_C + \cdots$$

where for component i

$$P_i = \frac{n_i RT}{V}$$

which is the partial pressure of i. This result is known as Dalton's Law of Partial Pressures.

If all gases in a chemical reaction are considered to be ideal gases, knowing the pressure, P_X, temperature, T_X and volume, V_X, for a component X that is a gas in the reaction allows one with the ideal gas equation of state to determine the number of moles of X by

$$n_X = \frac{P_X V_X}{RT_X}$$

Therefore the ideal gas equation provides a method to go from the lab world (P_X, V_X, T_X) to the mole world (n_X) in stoichiometry. If the number of moles of X (n_X) is known along with the volume (V_X) and the temperature (T_X) the partial pressure of gas X can be found from

$$P_X = n_X \left(\frac{RT_X}{V_X} \right)$$

Therefore the ideal gas equation of state also allows one to go from the mole world to the lab world. A similar expression exists for V_X if P_X, n_X and T_X are known or for T_X if P_X, V_X and n_X are known. A summary of all three types of conversions between the lab line and mole line is shown on the next page.

$$\begin{array}{ccccccc}
& xX & & xX & & xX & \\
\text{lab} & m_X & & V, [X] & & V_X, T_X, P_X & \\
& M \Updownarrow & \dfrac{1}{M} & V[X] \Updownarrow & \dfrac{n_X}{V} & \dfrac{P_X V_X}{RT_X} \Updownarrow & \dfrac{n_X RT_X}{V_X} \\
\text{mole} & n_X & & n_X & & n_X &
\end{array}$$

LR

To begin with, consider the Law of Combining Volumes which states that, for gaseous reactions with the temperature and pressure of all gases the same, the ratios of the volumes of all gases whether reactant gases or product gases in the reaction are expressible in terms small whole numbers. To see that stoichiometry gives this result and determines what the small whole numbers are consider the following general form of a gaseous reaction with the format shown below.

$$aA_{(g)} + bB_{(g)} \longrightarrow cC_{(g)} + dD_{(g)}$$

lab

mole

LR

Suppose that all gases have the same pressure, P, and are at the same temperature, T, and one wants to determine the volume of gas A that reacts and the volumes of gases C and D that are produced if one Liter of gas B is used in the reaction which is assumed to go to completion. This leads to the format shown on the next page.

Chapter 3: Stoichiometry | 99

$$aA_{(g)} + bB_{(g)} \longrightarrow cC_{(g)} + dD_{(g)}$$

lab   P,T,V =   P,T,1L   P,T,V =   P,T,V =

mole

LR

Since T and P are constant, it is convenient to let k=P/(RT) which has units of mol/L. With this one has n=kV and V=n/k. Using the format, the solution for this problem is shown below.

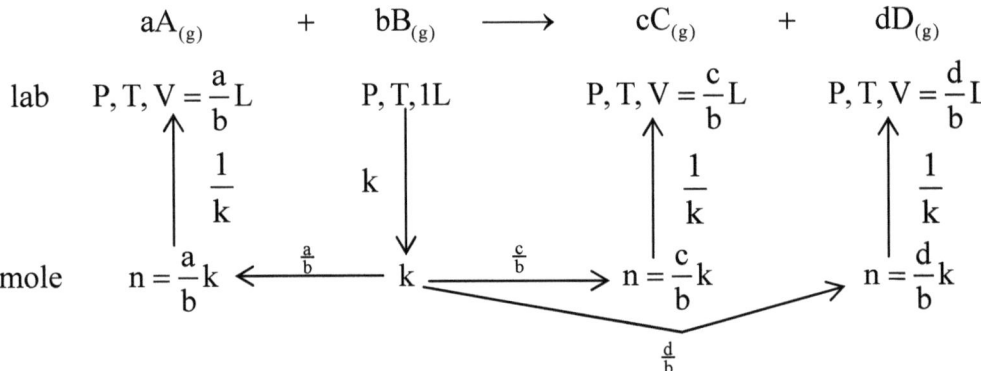

LR

One sees that these results lead to the Law of Combining Volumes. For example, the ratio of the product volumes is given by

$$\frac{V_D}{V_C} = \frac{\left(\frac{d}{b}\right)L}{\left(\frac{c}{b}\right)L} = \frac{d}{c}$$

which are small whole numbers. One also sees that these small whole numbers are actually the stoichiometric coefficients in the balanced reaction.

100 | Stoichiometry and Beyond

As another example, consider a reaction between hydrogen and nitrogen gases to form ammonia gas that takes place in a 5.00L container at 25.0°C. The objective is to determine the total pressure in the container after the maximum amount of reaction occurs when the initial partial pressures of hydrogen and nitrogen are 0.734atm and 0.196atm respectively. The temperature remains constant. The format that defines the limiting reagent is shown below.

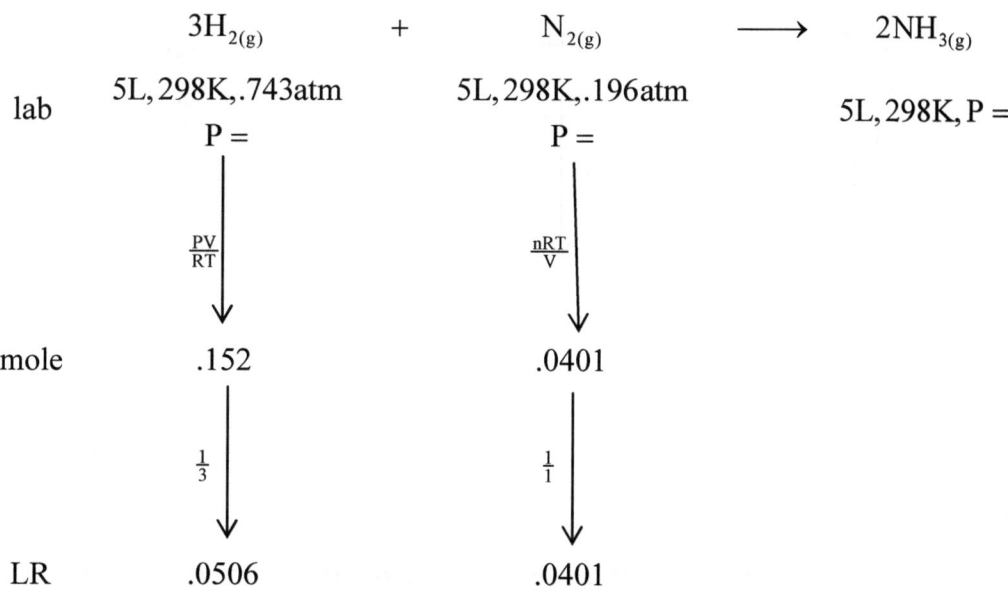

One sees that nitrogen is the limiting reagent so that 0.0401mol of reaction is the maximum amount that can occur. Using this on the LR line gives the partial pressures of hydrogen and ammonia as shown below

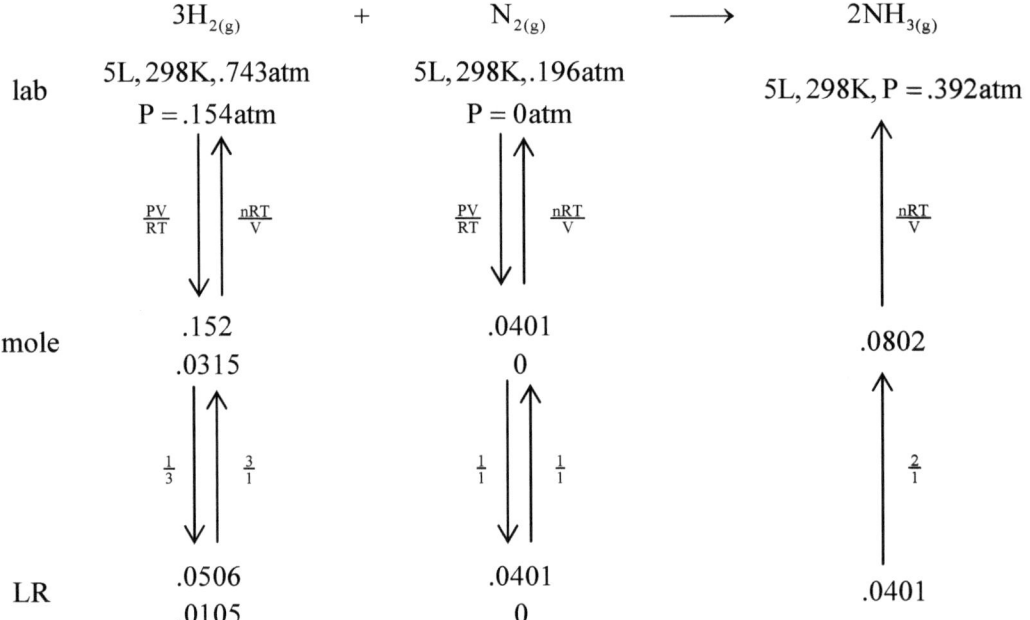

The lab line pressures are the partial pressures of each gas. Since the total pressure is the sum of the partial pressures, one has that

$$P_{tot} = P_{H_2} + P_{NH_3} = 0.154\,\text{atm} + 0.392\,\text{atm} = .546\,\text{atm}$$

where obviously the nitrogen gas has all reacted.

EXERCISES

The following exercises deal with the following reaction where A and B are the reactants and C and D are the products.

$$3A_{(g)} + 4B_{(g)} \longrightarrow 5C_{(g)} + 2D_{(g)}$$

Lab

Mole

LR

102 | Stoichiometry and Beyond

1) Determine the partial pressures of all gases and the total pressure when 2.25mol of A react with 2.90mol of B in a 15.0L container at 288K. (8.11atm)

2) Determine the partial pressures of all gases and the total pressure when gas A with P_A=0.469atm reacts with gas B with P_B=0.740atm in a 7.50L container at 298K. (1.21atm)

3) During the reaction the pressure of all gases is constant at 1.00atm and the temperature is constant at 298K. Determine what volume of A would needed and what volumes of C and D would be produced when 1.76L of gas B reacts. (Volume of C is 2.20L)

4) Before reaction gas A is at 298K, has P_A=1.15atm and is in a container of 5.34L and gas B is at 298K, has P_B=1.85atm and is in a container of 4.89L. A stopcock connects the containers. The stopcock is opened and the gases mix and react and the temperature increases to 326K. Determine the total pressure after the reaction. (1.97atm)

Chapter 3: Stoichiometry | 103

The following exercises deal with the following reaction where A and B are the reactants and C and D are the products. The molarity of B and the molar mass of D are also given.

$$\phantom{3A_{(g)} + }\ 2.75M \phantom{\ \longrightarrow 5C_{(g)} + }\ 57.9g/mol$$
$$3A_{(g)} + 4B_{(aq)} \longrightarrow 5C_{(g)} + 2D_{(s)}$$

Lab

Mole

LR

5) At a constant temperature of 298K in a container of 4.98L gas A has a pressure of 3.74atm. Determine the volume of B, the pressure of C and the mass of D after the reaction. (P_C=6.23atm)

6) The reaction occurs at a constant temperature of 298K in a container of 4.98L. Determine the final molarity of B, the mass of D and the total pressure when 54.0mL of B reacts with A with P_A=0.81atm. (Total pressure is 2.01atm)

CHAPTER 4
CHEMICAL EQUILIBRIUM

Introduction

In Chapter 3 it was shown how the three common features of all stoichiometry problems could be turned into answerable questions. Recording the answers in a common format in essence made all problems the same. The only things that changed were the chemical reaction, the measurements and the objectives. The methods of moving around in the format from the measurements to the objectives were very systematic. This chapter deals with chemical equilibrium, which also uses a balanced chemical reaction. There are also measurements and objectives so that the questions used in the previous chapter are also applicable here. The format will again be helpful in unifying this area of chemistry. Throughout this chapter it is assumed that only one reaction is occurring.

In Chapter 3 it was pointed out that many reactions do not proceed until something runs out and the percent yield was introduced to account for this. The percent yield, as will be shown, depends on the amounts of the reactants, which limit to some extent its utility. In this chapter we will find that it is possible to predict the final amounts of all reactants and products knowing only their initial amounts and the equilibrium constant.

In the next section we will define the condition of dynamic equilibrium for a chemical reaction. We will also define the general form of the equilibrium expression which involves an equilibrium constant and equilibrium concentrations of solutes and/or gases. This is followed with some general relationships for equilibrium constants. The reaction quotient is then introduced and its importance in determining which way a reaction proceeds is pointed out. This is followed by the introduction of a format that can be used for all problems in chemical equilibrium. This is done for both homogeneous and heterogeneous reactions. The general form of the equilibrium expression is also given and applied to reactions that give equilibrium expressions of second-degree polynomials. Following this a more general form of the equilibrium expression is introduced that can be entered into a calculator. It is also shown how the solver option on calculators can be used to find equilibrium quantities in a much more direct way and for a much larger class of reactions. The chapter closes with using Le Chatelier's Principle to predict how a disturbance affects the equilibrium state.

Dynamic Equilibrium and the Equilibrium Expression

For a chemical reaction of the form

$$\text{Reactants} \longrightarrow \text{Products}$$

and starting with only reactants present, stoichiometry predicts the amount of products that are formed assuming that the reaction goes to completion, i.e., a 100% reaction yield. However, many reactions do not go to completion so that the amounts of products are less than the predictions of stoichiometry. What is occurring in these situations is that the reverse reaction,

$$\text{Products} \longrightarrow \text{Reactants}$$

is also taking place. When the reaction is "over", the rates of the forward and reverse reactions are the same. Both reactions are still occurring but they "cancel" each other out so that no changes in the properties of the chemicals (concentrations, partial pressures, masses, etc.) take place. Reactions that have reached this point are depicted as

$$\text{Reactants} \rightleftarrows \text{Products}$$

and are said to be in a state of dynamic equilibrium or equilibrium for short.

Stoichiometry took these reactions into consideration by introducing the reaction yield. However, it turns out that the reaction yield depends on the initial amounts of reactants so that it is not a very general quantity to use for a reaction. A more general quantity for predicting the final amounts of products for a reaction is the equilibrium constant which does not depend on initial amounts. In defining the equilibrium constant in terms of properties of reactants and products, the general reaction consisting of reactants A and B and products C and D with corresponding stoichiometric coefficients a, b, c and d will be considered.

$$aA + bB \rightleftarrows cC + dD$$

If all compounds are solutes that are present in a solution with equilibrium molarities of $[A]_{eq}$, $[B]_{eq}$, $[C]_{eq}$ and $[D]_{eq}$, the equilibrium constant, K_c, is defined as

$$K_c = \frac{[C]_{eq}^c [D]_{eq}^d}{[A]_{eq}^a [B]_{eq}^b} \tag{1}$$

where the subscript "c" on K indicates it is defined in terms of concentrations (molarities). Equation (1) is referred to as the equilibrium expression. By convention, the products are in the numerator and the reactants are in the denominator. In general, the power of each concentration term is given by the stoichiometric coefficient of the corresponding chemical. It is very important to realize that the equilibrium constant does not depend on initial concentrations in that for any amounts of reactants and/or products initially present, the equilibrium concentrations will still satisfy Eq. (1). This is therefore much more general than the reaction yield. The equilibrium constant is a positive number and, depending on the reaction, can range from being very small to being very large. If it is very small, the reaction starting with only reactants proceeds to the product side to only a small extent. If it is very large, the reaction will proceed to the product side to a large extent.

It seems in Eq. (1) that K_c has units of molarity to the power of c+d-a-b but, in actuality, equilibrium constants are unit-less. The equilibrium constants actually come from thermodynamics and are defined in terms of dimensionless quantities called "activities". For solutes, an approximation to activity is given by Molarity divided by 1M which is unit-less. The value of 1M corresponds to the standard state of any solute. Rather than writing K_c properly as

$$K_c = \frac{\left(\frac{[C]_{eq}}{1M}\right)^c \left(\frac{[D]_{eq}}{1M}\right)^d}{\left(\frac{[A]_{eq}}{1M}\right)^a \left(\frac{[B]_{eq}}{1M}\right)^b}$$

it is easier to use Eq. (1) and just remember that the concentration terms have all been divided by 1M to give a unit-less expression. When writing down the expression for K_c, one must remember the following. The equilibrium expression only contains chemicals that change their concentration during the course of the reaction. In particular, the concentrations of pure solids and pure liquids are fixed and do not change in a reaction. Therefore they do not appear in the equilibrium expression. If, for example, A is a pure solid or pure liquid, the equilibrium expression would be

$$K_c = \frac{[C]_{eq}^c [D]_{eq}^d}{[B]_{eq}^b}$$

with the $[A]_{eq}$ not present. If the solvent appears in the reaction, it is in such great excess that its concentration does not appreciably change and is therefore not included in the equilibrium expression either. For example, if D is the solvent and all other compounds are solutes, one would have

$$K_c = \frac{[C]_{eq}^c}{[A]_{eq}^a [B]_{eq}^b}$$

where $[D]_{eq}$ does not appear. In general, the proper form of the equilibrium expression can be obtained by considering the subscripts on the compounds in the reaction. For a compound X, subscripts s and ℓ, i.e., $X_{(s)}$ or $X_{(\ell)}$ indicate pure solid or pure liquid respectively so that this compound would not be included in the equilibrium expression. If the compound is the solvent such as water, it is written as $H_2O_{(\ell)}$ and it would not be included either. If the compound is a gas, $X_{(g)}$ or a solute in water, $X_{(aq)}$, it is included in the equilibrium expression.

Another equilibrium constant is defined in terms of using partial pressures for all gases. This equilibrium constant, K_P, is especially important in thermodynamics. If A, B, C and D are all gases, then K_P is defined as

$$K_P = \frac{P_{Ceq}^c P_{Deq}^d}{P_{Aeq}^a P_{Beq}^b} \qquad (2)$$

which has the same form as Eq. (1) with [ ] replaced with P. K_P is again unit-less so that in actuality, each partial pressure was divided by the standard pressure in Eq. (2). Assuming all gases to be ideal gases gives

$$P_{Xeq} = \left(\frac{n_{Xeq}}{V}\right) RT = [X]_{eq} RT \qquad ,$$

and using this in Eq. (2) leads to

$$K_P = \frac{[C]_{eq}^c [D]_{eq}^d}{[A]_{eq}^a [B]_{eq}^b} (RT)^{c+d-a-b}$$

or

108 | Stoichiometry and Beyond

$$K_P = K_c(RT)^{c+d-a-b} \qquad (3)$$

which is a relation between the two equilibrium constants when all compounds in them are gases. For heterogeneous systems involving solutes in solution and gases, the more appropriate equilibrium constant is K_P, which uses molarities for solutes and partial pressures for gases. For example, if A and B are solutes, C is a solid and D is a gas, then

$$K_P = \frac{P_{Deq}^d}{[A]_{eq}^a [B]_{eq}^b} \quad.$$

For K_c, the [ ] symbol would be ambiguous since the volume of the gas does not correspond to the volume of the liquid. Finally, if there are no gases, K_c and K_P are the same.

General Relationships for Equilibrium Constants

Some general relationships for equilibrium constants will now be obtained using K_c.

1) When the reaction with the equilibrium constant K_c

$$aA + bB \rightleftarrows cC + dD$$

is turned around it becomes

$$cC + dD \rightleftarrows aA + bB$$

which has the equilibrium constant K_c'. However

$$K_c' = \frac{[A]_{eq}^a [B]_{eq}^b}{[C]_{eq}^c [D]_{eq}^d} = \frac{1}{\frac{[C]_{eq}^c [D]_{eq}^d}{[A]_{eq}^a [B]_{eq}^b}} = \frac{1}{K_c} \quad.$$

In general, if any reaction with equilibrium constant K_c is turned around, the equilibrium constant for the turned around reaction is $(K_c)^{-1}$.

2) This relation deals with multiplying a reaction by a number. If the reaction with the equilibrium constant K_c

$$aA + bB \rightleftharpoons cC + dD$$

is multiplied by a number, n, the reaction becomes

$$naA + nbB \rightleftharpoons ncC + ndD$$

which has the equilibrium constant K_c'. However one has

$$K_c' = \frac{[C]_{eq}^{nc}[D]_{eq}^{nd}}{[A]_{eq}^{na}[B]_{eq}^{nb}} = \left(\frac{[C]_{eq}^{c}[D]_{eq}^{d}}{[A]_{eq}^{a}[B]_{eq}^{b}}\right)^n = K_c^n$$

so that the new equilibrium constant is simply the old equilibrium constant to the power n. This holds for both positive and negative values of n.

3) This relation deals with the equilibrium constant that is obtained when reactions are combined. If one added n times the reaction with the equilibrium constant K_c

$$aA + bB \rightleftharpoons cC + dD$$

which becomes

$$naA + nbB \rightleftharpoons ncC + ndD$$

to m times the reaction

$$eE + fF \rightleftharpoons gG + hH$$

with equilibrium constant

$$K'_c = \frac{[G]^g_{eq}[H]^h_{eq}}{[E]^e_{eq}[F]^f_{eq}}$$

which becomes

$$meE + mfF \rightleftharpoons mgG + mhH$$

then the new reaction

$$naA + nbB + meE + mfF \rightleftharpoons ncC + ndD + mgG + mhH$$

has an equilibrium constant K''_c given by

$$K''_c = \frac{[C]^{cn}_{eq}[D]^{dn}_{eq}[G]^{gm}_{eq}[H]^{hm}_{eq}}{[A]^{an}_{eq}[B]^{bn}_{eq}[E]^{em}_{eq}[F]^{fm}_{eq}} = \left(\frac{[C]^c_{eq}[D]^d_{eq}}{[A]^a_{eq}[B]^b_{eq}}\right)^n \left(\frac{[G]^g_{eq}[H]^h_{eq}}{[E]^e_{eq}[F]^f_{eq}}\right)^m = K^n_c K'^m_c$$

which is just the product of the original equilibrium constants to their corresponding powers. If m=n=1,

$$K''_c = K_c K'_c$$

and if m=1 and n=-1, then

$$K''_c = \frac{K_c}{K'_c}$$

Analogous relations hold for the equilibrium constant K_P.

EXERCISES

1) For the following reactions, A and B are the reactant molecules and C and D are the product molecules along with the stoichiometric coefficients. For each reaction do the following.

   i) Write down when appropriate the expressions for both K_c and K_P.

   ii) Determine when appropriate K_P in terms of K_c and a power of RT.

iii) Determine K_P for $K_c=0.752$ and $T=295K$.

a) $3A_{(aq)} + B_{(aq)} \rightleftharpoons 2C_{(aq)} + 2D_{(aq)}$

b) $2A_{(s)} + B_{(g)} \rightleftharpoons 3C_{(\ell)} + 2D_{(g)}$

c) $A_{(\ell)} + 2B_{(aq)} \rightleftharpoons 2C_{(aq)} + 3D_{(g)}$

d) $3A_{(s)} + 2B_{(g)} \rightleftharpoons 2C_{(aq)} + 2D_{(g)}$

e) $4A_{(aq)} + B_{(g)} \rightleftharpoons 2C_{(s)} + 3D_{(aq)}$

f) $A_{(s)} + 2B_{(\ell)} \rightleftharpoons 2C_{(s)} + 2D_{(g)}$

2) This question deals with two reactions (1) and (2) and their equilibrium constants K_1 and K_2 given below.

$$\text{Rx1} \quad R_1 \rightleftharpoons P_1 \quad K_1 = 2.5$$
$$\text{Rx2} \quad R_2 \rightleftharpoons P_2 \quad K_2 = 4.5$$

Determine the equilibrium constant for the following situations.

a) -Rx1

b) Rx1+Rx2

c) 2Rx1-Rx2

d) (1/2)Rx1+3Rx2

e) -(1/3)Rx1-(1/2)Rx2

f) 3Rx2-4Rx1

g) -(3/2)Rx2+(1/3)Rx1

The Reaction Quotient

Another quantity that will be found to be very useful is called the reaction quotient (Q). The reaction quotient (Q_c or Q_P) has the same form as the equilibrium constant (K_c or K_P) but instead of equilibrium concentrations or partial pressures, it is defined in terms of the actual concentrations or partial pressures at any stage of the reaction. For example Q_c corresponding to Eq. (1) is defined as

$$Q_c = \frac{[C]^c[D]^d}{[A]^a[B]^b} \tag{4}$$

and Q_P corresponding to Eq. (2) is defined as

$$Q_P = \frac{P_C^c P_D^d}{P_A^a P_B^b} \tag{5}$$

where the subscripts "eq" are not present indicating that the quantities are defined at any stage of the reaction. The usefulness of the reaction quotient becomes apparent when it is compared to the equilibrium constant. When Q (either Q_c or Q_P) is compared to K (either K_c or K_P), three results are possible. One result could be that

$$Q < K$$

i.e., the reaction quotient is less that the equilibrium constant. Since the reaction will proceed until it reaches equilibrium, the ratio of the numerator involving product concentrations or partial pressures to denominator involving reactant concentrations or partial pressures is too small. Therefore, as the reaction proceeds to equilibrium, the numerator must increase and the denominator must decrease. For this to occur, the concentrations or partial pressures of the products must increase so that the reaction must go to the product side. Another result could be that

$$Q > K$$

so that the numerator involving products divided by the denominator involving reactants is too large. Therefore, as the reaction proceeds to equilibrium, the numerator must decrease and the denominator must increase. For this to occur, the concentrations or partial pressures of the products must decrease so that the reaction must go to the reactant side. The final possibility is that

$$Q = K$$

which indicates that the reaction is at equilibrium and no shift will occur. Indicating with an arrow which side the reaction will proceed ($\longrightarrow$ product side, $\longleftarrow$ reactant side or $\rightleftarrows$ no shift, i.e. equilibrium), this is summarized in the following table.

condition	shift
Q<K	$\longrightarrow$
Q>K	$\longleftarrow$
Q=K	$\rightleftarrows$

In the next section, it will be seen that it is very helpful to know, from initial measurements, which way the reaction proceeds in order to set up an unambiguous solution for the equilibrium concentrations or partial pressures. Since the reaction quotient is obviously defined for the initial state of the reaction, one can always determine which way the reaction will proceed.

The Equilibrium Format

A general treatment of the equilibrium state of a chemical reaction will now be presented. The three questions and the format used in the solution of problems in stoichiometry will still apply in this area. Three cases will be considered. The first case will be homogeneous reactions described in terms of concentrations. Homogeneous means that all compounds in the equilibrium expression are in the same phase. The second case will be homogeneous reactions described in terms of partial pressures. The final case will involve heterogeneous reactions. Heterogeneous means that not all compounds in the equilibrium expression are in the same phase.

The first question is "What is the Chemistry?". It is assumed that all reactants and products are solutes. For reactants A and B and products C and D with corresponding stoichiometric coefficients a, b, c and d, the answer to the first question gives the following format.

$$aA_{(aq)} + bB_{(aq)} \rightleftharpoons cC_{(aq)} + dD_{(aq)}$$

lab   in
       eq

mole   in
       eq

LR

The lab and mole lines now have an initial line (in) and an equilibrium line (eq), which corresponds to initial quantities and equilibrium quantities.

Thee second question is "What are the measurements?". It is assumed that the initial concentrations (designated with a subscript "o") are given for all reactants and products and also their corresponding volumes, which gives the format below.

Chapter 4: Chemical Equilibrium | 115

$$aA_{(aq)} + bB_{(aq)} \rightleftharpoons cC_{(aq)} + dD_{(aq)}$$

lab	in	$[A]_o, V_A$	$[B]_o, V_B$	$[C]_o, V_C$	$[D]_o, V_D$
	eq				

mole	in				
	eq				

LR

If the compounds are initially all present in the same solution then $V_A=V_B=V_C=V_D=V$ where V is the final volume. If solutions of the compounds are mixed to give a final solution then $V=V_A+V_B+V_C+V_D$ where again V is the final volume. Finally the answer to the third question "What are the objectives?" is to determine all equilibrium concentrations, which gives the following format.

$$aA_{(aq)} + bB_{(aq)} \rightleftharpoons cC_{(aq)} + dD_{(aq)}$$

lab	in	$[A]_o, V_A$	$[B]_o, V_B$	$[C]_o, V_C$	$[D]_o, V_D$
	eq	$[A]_{eq} =$	$[B]_{eq} =$	$[C]_{eq} =$	$[D]_{eq} =$

mole	in				
	eq				

LR

To solve the problem we can go to the initial mole line by multiplying the initial molarities by the initial volumes as is shown in the format below.

$$aA_{(aq)} + bB_{(aq)} \rightleftharpoons cC_{(aq)} + dD_{(aq)}$$

		$aA_{(aq)}$	$bB_{(aq)}$	$cC_{(aq)}$	$dD_{(aq)}$
lab	in	$[A]_o, V_A$	$[B]_o, V_B$	$[C]_o, V_C$	$[D]_o, V_D$
	eq	$[A]_{eq} =$	$[B]_{eq} =$	$[C]_{eq} =$	$[D]_{eq} =$
mole	in	n_A^{in}	n_B^{in}	n_C^{in}	n_D^{in}
	eq				
LR					

At equilibrium, we know that there is a unique number of moles for each reactant and product as well as a unique number of moles of reaction that has occurred which is entered into the format below.

$$aA_{(aq)} + bB_{(aq)} \rightleftharpoons cC_{(aq)} + dD_{(aq)}$$

		$aA_{(aq)}$	$bB_{(aq)}$	$cC_{(aq)}$	$dD_{(aq)}$
lab	in	$[A]_o, V_A$	$[B]_o, V_B$	$[C]_o, V_C$	$[D]_o, V_D$
	eq	$[A]_{eq} =$	$[B]_{eq} =$	$[C]_{eq} =$	$[D]_{eq} =$
mole	in	n_A^{in}	n_B^{in}	n_C^{in}	n_D^{in}
	eq	$n_A^{eq} =$	$n_B^{eq} =$	$n_C^{eq} =$	$n_D^{eq} =$
LR		n_{Rx}	n_{Rx}	n_{Rx}	n_{Rx}

An important objective here is to obtain a general form for the equilibrium expression that contains initial concentrations. If the reaction proceeds to the product side, then for reactant A $n_A^{eq} < n_A^{in}$ and for product D $n_D^{eq} > n_D^{in}$. However if the reaction proceeds to the reactant side, then for reactant A $n_A^{eq} > n_A^{in}$ and for product D $n_D^{eq} < n_D^{in}$. If n_{Rx} is required to be positive then if the reaction goes to the product side $n_A^{eq} = n_A^{in} - an_{Rx}$ and $n_D^{eq} = n_D^{in} + dn_{Rx}$ or if the reaction goes to the reactant side $n_A^{eq} = n_A^{in} + an_{Rx}$ and $n_D^{eq} = n_D^{in} - dn_{Rx}$. Rather than have this situation a general form of the equilibrium expression can be obtained by always using minus signs on the reactant side and plus signs on the product side so that $n_A^{eq} = n_A^{in} - an_{Rx}$ and $n_D^{eq} = n_D^{in} + dn_{Rx}$. With this convention n_{Rx} will be positive if the reaction proceeds to the product side and negative if it proceeds to the reactant side. We will show that the sign of n_{Rx} is easily determined by comparing the reaction quotient to the equilibrium constant. Using this convention gives the format shown below.

		$aA_{(aq)}$	+	$bB_{(aq)}$	$\rightleftharpoons$	$cC_{(aq)}$	+	$dD_{(aq)}$
lab	in	$[A]_o, V_A$		$[B]_o, V_B$		$[C]_o, V_C$		$[D]_o, V_D$
	eq	$[A]_{eq} =$		$[B]_{eq} =$		$[C]_{eq} =$		$[D]_{eq} =$
mole	in	n_A^{in}		n_B^{in}		n_C^{in}		n_D^{in}
	eq	$n_A^{eq} = n_A^{in} - an_{rx}$		$n_B^{eq} = n_B^{in} - bn_{rx}$		$n_C^{eq} = n_C^{in} + cn_{rx}$		$n_D^{eq} = n_D^{in} + dn_{rx}$
		$\uparrow \frac{a}{1}$		$\uparrow \frac{b}{1}$		$\uparrow \frac{c}{1}$		$\uparrow \frac{d}{1}$
LR		n_{Rx}		n_{Rx}		n_{Rx}		n_{Rx}

Now by dividing the equilibrium number of moles by the final volume we can go to the lab equilibrium line, which is shown in the following format.

		$aA_{(aq)}$	+	$bB_{(aq)}$	$\rightleftharpoons$	$cC_{(aq)}$	+	$dD_{(aq)}$
lab	in	$[A]_o, V_A$		$[B]_o, V_B$		$[C]_o, V_C$		$[D]_o, V_D$
	eq	$[A]_{eq} = [A]_o' - ax$		$[B]_{eq} = [B]_o' - bx$		$[C]_{eq} = [C]_o' + cx$		$[D]_{eq} = [D]_o' + dx$
mole	in	n_A^{in}		n_B^{in}		n_C^{in}		n_D^{in}
	eq	$n_A^{eq} = n_A^{in} - an_{rx}$		$n_B^{eq} = n_B^{in} - bn_{rx}$		$n_C^{eq} = n_C^{in} + cn_{rx}$		$n_D^{eq} = n_D^{in} + dn_{rx}$
		$\uparrow \frac{a}{1}$		$\uparrow \frac{b}{1}$		$\uparrow \frac{c}{1}$		$\uparrow \frac{d}{1}$
LR		n_{Rx}		n_{Rx}		n_{Rx}		n_{Rx}

For brevity x above is defined as

$$x = [n_{Rx}] = \frac{n_{Rx}}{V}$$

where V is the final volume. Notice that the initial concentrations used on the lab equilibrium line are not necessarily the ones used on the lab initial line as indicated by using a prime on the equilibrium line. If all solutes that were initially present were in the same volume then if A were initially present $[A]'_o = [A]_o$. If all solutes that were initially present were in different containers with corresponding volumes then $V=V_A + V_B + V_C + V_D$ and if A were initially present $[A]'_o = n_A^{in}/V = [A]_o V_A/V$. Note that this is just a dilution of A. Other reactants or products are treated in a similar fashion. One notices that in the format the mole line is not really needed to define the equilibrium concentrations and a more compact format shown below can be used.

		$aA_{(aq)}$	+	$bB_{(aq)}$	$\rightleftharpoons$	$cC_{(aq)}$	+	$dD_{(aq)}$
lab	in	$[A]_o, V_A$		$[B]_o, V_B$		$[C]_o, V_C$		$[D]_o, V_D$
	eq	$[A]_{eq} = [A]'_o - ax$		$[B]_{eq} = [B]'_o - bx$		$[C]_{eq} = [C]'_o + cx$		$[D]_{eq} = [D]'_o + dx$
		$\frac{1}{a}\downarrow$		$\frac{1}{b}\downarrow$		$\frac{1}{c}\downarrow$		$\frac{1}{d}\downarrow$
LC		$\left[n_{Rx}^{in}\right]$		$\left[n_{Rx}^{in}\right]$		$\left[n_{Rx}^{in}\right]$		$\left[n_{Rx}^{in}\right]$

In the shortened format the LR line has been replaced with the LC line, which will determine which compound is the limiting reagent on each side. The values on this line are found by dividing the initial concentration of a compound on the lab eq line by its corresponding stoichiometric coefficient (for A it would be $[A]'_o V_A/aV$). This allows one to determine the maximum amount of reaction that can occur which defines the bounds for where x must be. This is sometimes helpful in finding the value of x. Using the information on the lab equilibrium line gives a general equation for the equilibrium expression as

$$K_c = \frac{([C]'_o + cx)^c ([D]'_o + dx)^d}{([A]'_o - ax)^a ([B]'_o - bx)^b} \quad . \tag{6}$$

There are, given the initial concentrations, two unknowns in the equilibrium expression, which are K_c and x. This suggests two general types of problems. One type involves knowing the initial concentrations and being able to determine x and one can then determine the equilibrium constant. The other type involves knowing the initial concentrations and the equilibrium constant so that one can determine x and the equilibrium concentrations. Rearranging the equilibrium expression gives

$$K_c ([A]'_o - ax)^a ([B]'_o - bx)^b - ([C]'_o + cx)^c ([D]'_o + dx)^d = 0 \tag{7}$$

and each term can be re-expressed as a polynomial. After the polynomials are multiplied and all terms with a common power of x are grouped together one obtains a polynomial of x. The order of this polynomial is a+b if a+b is greater or equal to c+d or c+d if c+d is greater or equal to a+b. The values of x that satisfy this polynomial being equal to zero are called the roots of the polynomial. Since many polynomials of x have both positive and negative roots and one has only one of the roots that give a meaningful (physical) solution of the problem one must know whether x should be positive or negative. This is done by evaluating the reaction quotient given by

$$Q_c = \frac{([C]'_o)^c ([D]'_o)^d}{([A]'_o)^a ([B]'_o)^b}$$

where again the final volume is needed to evaluate Q_c. If $Q_c > K_c$ the reaction proceeds to the reactant side and x<0. If $Q_c < K_c$ the reaction proceeds to the product side and x>0. Finally if $Q_c = K_c$ the reaction is at equilibrium and x=0.

Formulas for finding the roots of polynomials can be quite complicated even for third order (x^3) polynomials. However, if both $a+b \leq 2$ and $c+d \leq 2$, one is either dealing with a linear equation or a quadratic polynomial of x. For a linear equation of the form

$$a_0 + a_1 x = 0$$

the solution of x is

$$x = \frac{a_0}{a_1} .$$

For a quadratic polynomial of x of the form

$$a_0 + a_1 x + a_2 x^2 = 0$$

x has two roots given by the quadratic formula

$$x_\pm = \frac{-a_1 \pm \sqrt{a_1^2 - 4 a_0 a_2}}{2 a_2}$$

where the root x_+ is obtained by using a + in front of the radical and the root x_- is obtained by using a − in front of the radical. As a specific case that gives equilibrium expressions in the form of quadratic polynomials we will let a=b=c=d=1. We will further assume that all solutes on the in and eq lab line occupy the same volume V which is not needed in the format. This gives the following format.

$$
\begin{array}{cccccccc}
 & & A_{(aq)} & + & B_{(aq)} & \rightleftharpoons & C_{(aq)} & + & D_{(aq)} \\
\text{lab} & \text{in} & [A]_o & & [B]_o & & [C]_o & & [D]_o \\
 & \text{eq} & [A]_{eq}=[A]_o-x & & [B]_{eq}=[B]_o-x & & [C]_{eq}=[C]_o+x & & [D]_{eq}=[D]_o+x \\
 & & \downarrow \frac{1}{1} & & \downarrow \frac{1}{1} & & \downarrow \frac{1}{1} & & \downarrow \frac{1}{1} \\
 & \text{LC} & [A]_o & & [B]_o & & [C]_o & & [D]_o
\end{array}
$$

The equilibrium expression is given by

$$K_c = \frac{([C]_o + x)([D]_o + x)}{([A]_o - x)([B]_o - x)}.$$

Rearranging gives

$$K_c([A]_o - x)([B]_o - x) - ([C]_o + x)([D]_o + x) = 0.$$

Multiplying out the terms gives

$$K_c[A]_o[B]_o - K_c[A]_o x - K_c[B]_o x + K_c x^2 - [C]_o[D]_o - [C]_o x - [D]_o x - x^2 = 0$$

and grouping terms with common powers of x finally gives

$$K_c[A]_o[B]_o - [C]_o[D]_o - (K_c[A]_o + K_c[B]_o + [C]_o + [D]_o)x + (K_c - 1)x^2 = 0 \quad . \quad (8)$$

This illustrates an (in my opinion) "unpleasant aspect" of solving equilibrium problems, which is a lot of manipulation of equations to get a solvable form. Now suppose that the measurements are $[A]_o=1.0M$, $[B]_o=1.5M$, $[C]_o=1.3M$, $[D]_o=0.25M$ and $[D]_{eq}=0.65M$. Suppose also that the objective is to determine the equilibrium constant. This gives the format below.

		$A_{(aq)}$	+	$B_{(aq)}$	⇌	$C_{(aq)}$	+	$D_{(aq)}$
lab	in	1.0		1.5		1.3		.25
	eq	$[A]_{eq} = 1 - x$		$[B]_{eq} = 1.5 - x$		$[C]_{eq} = 1.3 + x$		$.65 = .25 + x$
	LC							

As one sees on the eq line for D x=.65-.25=.4. Then

$$K_c = \frac{(1.3+.4)(.65)}{(1-.4)(1.5-.4)} = \frac{(1.7)(.65)}{(.6)(1.1)} = 1.7$$

which solves the problem. Notice how the format helped in the solution.

Suppose now that the measurements are K_c=1.7, $[A]_o$=.25M, $[B]_o$=.57M, $[C]_o$=1.2M and $[D]_o$=.85M the objectives are to determine all equilibrium concentrations. The format with the measurements and objectives is given below.

		$A_{(aq)}$	+	$B_{(aq)}$	⇌	$C_{(aq)}$	+	$D_{(aq)}$
lab	in	.25		.57		1.2		.85
	eq	$[A]_{eq} = .25 - x$		$[B]_{eq} = .57 - x$		$[C]_{eq} = 1.2 + x$		$[D]_{eq} = .85 + x$
	LC	.25		.57		1.2		.85

In order to determine whether x is positive or negative one needs to evaluate the reaction quotient, which is

$$Q_c = \frac{(1.2)(.85)}{(.25)(.57)} = 7.2 > K_c = 1.7 \ .$$

Since $Q_c>K_c$ the reaction proceeds to the reactant side so that x<0. Using the initial concentrations and K_c in Eq. (8) gives

$$1.7(.25)(.57) - (1.2)(.85) - (1.7(.25+.57) + 1.2 + .85)x + (1.7-1)x^2 = 0$$

or

$$-.78 - 3.44x + .7x^2 = 0.$$

Using the quadratic formula to find the roots gives

$$x_\pm = \frac{-(-3.44) \pm \sqrt{(3.44)^2 - 4(-.78)(.7)}}{2(.7)}$$

or

$$x_\pm = \frac{3.44 \pm 3.74}{1.4}.$$

In order to get a negative root one must use the - sign which gives x=-0.22. Using this value of x gives the equilibrium concentrations shown in the format below.

		$A_{(aq)}$ +	$B_{(aq)}$	⇌	$C_{(aq)}$ +	$D_{(aq)}$
lab	in	.25	.57		1.2	.85
	eq	$[A]_{eq} = .47$	$[B]_{eq} = .79$		$[C]_{eq} = .98$	$[D]_{eq} = .63$
	LC	.25	.57		1.2	.85

For brevity the unit of M for the concentrations are not included in the format. One notices that in this problem the quantities on the LC line were not used and in many problems this will be the case. It will be shown however that this line can be helpful.

Another class of reactions that are used extensively especially for solutions of acids and bases have a=b=c=d=1 where A, C and D are solutes but B is a solvent. This results in the format below.

		$A_{(aq)}$ +	$B_{(\ell)}$	⇌	$C_{(aq)}$ +	$D_{(aq)}$
lab	in	$[A]_o$	xs		$[C]_o$	$[D]_o$
	eq	$[A]_{eq} = [A]_o - x$			$[C]_{eq} = [C]_o + x$	$[D]_{eq} = [D]_o + x$
	LC	$[A]_o$			$[C]_o$	$[D]_o$

The equilibrium expression is

$$K_c = \frac{([C]_o + x)([D]_o + x)}{([A]_o - x)}$$

which is rearranged to give

$$K_c[A]_o - [C]_o[D]_o - (K_c + [C]_o + [D]_o)x - x^2 = 0 \qquad (9)$$

Suppose that the measurements are $K_c = 1.4 \times 10^{-7}$ and only A is present initially with $[A]_o = .75M$ and the objectives are to determine the equilibrium concentrations. This gives the following format.

$$
\begin{array}{cccccccc}
 & & A_{(aq)} & + & B_{(\ell)} & \rightleftarrows & C_{(aq)} & + & D_{(aq)} \\
\text{lab} & \text{in} & .75 & & \text{xs} & & 0 & & 0 \\
 & \text{eq} & [A]_{eq} = .75 - x & & & & [C]_{eq} = x & & [D]_{eq} = x \\
\end{array}
$$

LC

Using this information in Eq. (9) gives

$$1.05 \times 10^{-7} - 1.4 \times 10^{-7} x - x^2 = 0$$

and using the quadratic formula gives

$$x_{\pm} = \frac{1.4 \times 10^{-7} \pm \sqrt{(1.4 \times 10^{-7})^2 - 4(1.05 \times 10^{-7})(-1)}}{2(-1)}$$

or

$$x_{\pm} = -\frac{1.4 \times 10^{-7} \pm 6.48 \times 10^{-4}}{2}.$$

Since there are initially no products x must be greater than zero and using the − sign gives $x = 3.2 \times 10^{-4}$. Using this gives the equilibrium concentrations in the format below.

$$
\begin{array}{cccccccc}
 & & A_{(aq)} & + & B_{(\ell)} & \rightleftarrows & C_{(aq)} & + & D_{(aq)} \\
\text{lab} & \text{in} & .75 & & \text{xs} & & 0 & & 0 \\
 & \text{eq} & [A]_{eq} = .75 & & & & [C]_{eq} = 3.2 \times 10^{-4} & & [D]_{eq} = 3.2 \times 10^{-4} \\
\end{array}
$$

LC

Notice that the initial concentration of A has not significantly changed which is what one would expect for such a small K_c. If one assumes that $[A]_{eq}=[A]_o$ and only A is initially present in the equilibrium expression then

$$K_c = \frac{x^2}{[A]_o}$$

and

$$x = \sqrt{[A]_o K_c} \quad .$$

Using the measurements gives $x = \sqrt{.75(1.4 \times 10^{-7})} = 3.2 \times 10^{-4}$ as was previously obtained. In order to demonstrate something of importance suppose the measurements are $K_c=1.4 \times 10^{-7}$, $[A]_o=0M$, $[C]_o=.75M$ and $[D]_o=.75M$ and the objectives are to determine the equilibrium concentrations. Using this gives the following format.

		$A_{(aq)}$	+	$B_{(\ell)}$	$\rightleftharpoons$	$C_{(aq)}$	+	$D_{(aq)}$
lab	in	0		xs		.75		.75
	eq	$[A]_{eq} = -x$				$[C]_{eq} = .75 + x$		$[D]_{eq} = .75 + x$

LC

Using the information in Eq. (9) gives

$$.5625 + 1.5x + x^2 = 0$$

and using the quadratic formula gives

$$x_{\pm} = \frac{-1.5 \pm \sqrt{(1.5)^2 - 4(.5625)(1)}}{2(1)}$$

or

$$x_{\pm} = \frac{-1.5 \pm 0}{2} \quad .$$

which gives $x=-.75$. Using this gives the equilibrium concentrations in the format below.

Chapter 4: Chemical Equilibrium | 125

$$\begin{array}{cccccccc}
 & & A_{(aq)} & + & B_{(\ell)} & \rightleftarrows & C_{(aq)} & + & D_{(aq)} \\
\text{lab} & \text{in} & 0 & & \text{xs} & & .75 & & .75 \\
 & \text{eq} & [A]_{eq}=.75 & & & & [C]_{eq}=0 & & [D]_{eq}=0
\end{array}$$

LC

The zero equilibrium concentrations for C and D are not actually correct. The correct values could be obtained by using more accurate (increase the precision) numbers on the calculator. They can also be found from the equilibrium expression which is given by

$$1.4 \times 10^{-7} = \frac{[C]_{eq}[D]_{eq}}{.75}$$.

Since the equilibrium concentrations of C and D are the same taking the square root of .75 times 1.4×10^{-7} gives the same equilibrium concentrations as in the previous example. The important thing to remember from this example and the previous example is that as long as there is the same amount of material (.75M reactants or .75M products) the reaction will reach the same equilibrium state from whatever initial state it begins in. One can define the percent reaction in a general way as

$$\%Y = \frac{n_{Rx}}{n_{Rxtheor}} 100 = \frac{\frac{n_{Rx}}{V}}{\frac{n_{Rxtheor}}{V}} 100 = \frac{x}{x_{theor}} 100$$.

For either reaction the x_{theor} would be the entry on the LC line which is .75 and x is 3.2×10^{-4}. Therefore the percent yield is

$$\%Y = \frac{3.2 \times 10^{-4}}{.75} 100 = 0.043\%$$

Consider now the first example with $[A]_o=1.1M$. This gives the following format.

		$A_{(aq)}$	+	$B_{(\ell)}$	$\rightleftharpoons$	$C_{(aq)}$	+	$D_{(aq)}$
lab	in	1.1		xs		0		0
	eq	$[A]_{eq} = 1.1 - x$				$[C]_{eq} = x$		$[D]_{eq} = x$

LC

Using the information in Eq. (9) gives

$$1.54 \times 10^{-7} - 1.4 \times 10^{-7} x - x^2 = 0$$

and using the quadratic formula gives

$$x_{\pm} = \frac{1.4 \times 10^{-7} \pm \sqrt{(1.4 \times 10^{-7})^2 - 4(1.54 \times 10^{-7})(-1)}}{2(-1)}$$

or

$$x_{\pm} = -\frac{1.4 \times 10^{-7} \pm 7.85 \times 10^{-4}}{2}.$$

Using the $-$ sign gives $x = 3.9 \times 10^{-4}$. Using this gives the equilibrium concentrations in the format below.

		$A_{(aq)}$	+	$B_{(\ell)}$	$\rightleftharpoons$	$C_{(aq)}$	+	$D_{(aq)}$
lab	in	1.1		xs		0		0
	eq	$[A]_{eq} = 1.1$				$[C]_{eq} = 3.9 \times 10^{-4}$		$[D]_{eq} = 3.9 \times 10^{-4}$

LC

For the reaction yield one obtains

$$\%Y = \frac{3.9 \times 10^{-4}}{1.1} 100 = 0.035\%$$

which is different than the previous result. This example illustrates the statement that the percent yield is not as general a property as the equilibrium constant for a reaction because it depends on initial amounts.

As a final example that illustrates the usefulness of the LC line let the measurements be $K_c=2.7 \times 10^{-10}$ and $[A]_o=0M$, $[C]_o=.74M$ and $[D]_o=1.3M$. Using this gives the format below.

		$A_{(aq)}$	+	$B_{(\ell)}$	$\rightleftarrows$	$C_{(aq)}$	+	$D_{(aq)}$
lab	in	0		xs		.74		1.3
	eq	$[A]_{eq} = -x$				$[C]_{eq} = .74 + x$		$[D]_{eq} = 1.3 + x$
	LC					.74		1.3

Again we would obtain a zero equilibrium concentration for either C or D and have to obtain it from the equilibrium expression. However on the LC line there is .74 for C and 1.3 for D. Therefore C must be the limiting reagent. Since it is immaterial what the initial state is as long as the total amount of material is the same we could start the reaction on the LC line with .74 for A, 0 for C and 1.3-.74=.56 for D. With these quantities on the LC line the initial concentration would be .74M for A, 0M for C and .56M for D, which is shown in the format below.

		$A_{(aq)}$	+	$B_{(\ell)}$	$\rightleftarrows$	$C_{(aq)}$	+	$D_{(aq)}$
lab	in	.74		xs		0		.56
	eq	$[A]_{eq} = .74 - x$				$[C]_{eq} = x$		$[D]_{eq} = .56 + x$
		$\uparrow \frac{1}{1}$				$\uparrow \frac{1}{1}$		$\uparrow \frac{1}{1}$
	LC	.74				0		.56

Since K_c is so small we would expect that x would be very small and not significantly change the initial concentrations for A and D. Using this gives the following format.

		$A_{(aq)}$	+	$B_{(\ell)}$	$\rightleftarrows$	$C_{(aq)}$	+	$D_{(aq)}$
lab	in	.74		xs		0		.56
	eq	$[A]_{eq} = .74$				$[C]_{eq}$		$[D]_{eq} = .56$
		$\uparrow \frac{1}{1}$				$\uparrow \frac{1}{1}$		$\uparrow \frac{1}{1}$
	LC	.74				0		.56

The equilibrium concentration for C is found from the equilibrium expression, which becomes

$$2.7 \times 10^{-10} = \frac{[C]_{eq} \cdot .56}{.74}$$

so that $[C]_{eq}=2.7\times10^{-10}(.74)/.56=3.6\times10^{-10}$M. Note that you can always check to see if your assumptions are correct by putting the equilibrium concentrations into the equilibrium expression and seeing if you obtain the correct equilibrium constant.

The treatment when all compounds are gases in terms of partial pressures is entirely analogous to the treatment for concentrations. It is assumed that the reaction takes place in a container at constant temperature so that V and T of all gases are the same. It is a good exercise to go through the complete analysis and obtain the following format.

		$aA_{(g)}$	+	$bB_{(g)}$	$\rightleftarrows$	$cC_{(g)}$	+	$dD_{(g)}$
lab	in	P_A^o		P_B^o		P_C^o		P_D^o
	eq	$P_A^{eq}=P_A^o-ax$		$P_B^{eq}=P_B^o-bx$		$P_C^{eq}=P_C^o+cx$		$P_D^{eq}=P_D^o+dx$
		$\frac{1}{a}\downarrow$		$\frac{1}{b}\downarrow$		$\frac{1}{c}\downarrow$		$\frac{1}{d}\downarrow$
LC		P_{Rx}^o		P_{Rx}^o		P_{Rx}^o		P_{Rx}^o

The quantity, x, has units of pressure and is defined by

$$x = P_{Rx} = \frac{n_{Rx}RT}{V} \; .$$

As one can see the only change is that molarities are replaced with partial pressures. The LC line still tells you the limiting reagent on both sides of the reaction. For A $P_{Rx}^o = P_A^o/a$. Using the information on the lab equilibrium line gives a general equation for the equilibrium expression as

$$K_P = \frac{\left(P_C^o+cx\right)^c \left(P_D^o+dx\right)^d}{\left(P_A^o-ax\right)^a \left(P^o-bx\right)^b} \; . \tag{10}$$

There are, given the initial partial pressures, two unknowns in the equilibrium expression which are K_c and x. This suggests two general types of problems. One type involves knowing the initial partial pressures and being able to determine x and one can then determine the equilibrium constant. The other type involves knowing the initial partial pressures and the equilibrium constant so that one can determine x and the equilibrium partial pressures. Rearranging the equilibrium expression gives

$$K_P\left(P_A^o-ax\right)^a \left(P_B^o-bx\right)^b - \left(P_C^o+cx\right)^c \left(P_D^o+dx\right)^d = 0 \tag{11}$$

which is a polynomial of x. The reaction quotient for initial partial pressures is given by

$$Q_P = \frac{\left(P_C^o\right)^c \left(P_D^o\right)^d}{\left(P_A^o\right)^a \left(P_B^o\right)^b} \quad . \tag{12}$$

Note that all equations have the same structure so that the previous examples could be re done by changing K_c to K_P and interpreting the initial molarities as initial partial pressures.

The treatment for a heterogeneous reaction is similar to the previous treatments. In particular, assume that A, B and C are solutes and D is a gas and that the solutes are in the volume V and the gas occupies a volume V_g. With initial molarities of the solutes of $[A]_o$, $[B]_o$ and $[C]_o$ and an initial partial pressure of D of P_D^o, the analysis for the reaction occurring at temperature T is shown below.

		$aA_{(aq)}$	+	$bB_{(aq)}$	⇌	$cC_{(aq)}$	+	$dD_{(g)}$
lab	in	$[A]_o$		$[B]_o$		$[C]_o$		P_D^o
	eq	$[A]_{eq} = [A]_o - ax$		$[B]_{eq} = [B]_o - bx$		$[C]_{eq} = [C]_o + cx$		$P_D^{eq} = P_D^o + dP_{Rx}^{eq}$
		$\frac{1}{a} \downarrow$		$\frac{1}{b} \downarrow$		$\frac{1}{c} \downarrow$		$\frac{1}{d} \downarrow$
LC		$\left[n_{Rx}^{in}\right]$		$\left[n_{Rx}^{in}\right]$		$\left[n_{Rx}^{in}\right]$		P_{Rx}^o

The problem is that both $x=[n_{Rx}]=n_{Rx}/V$ and $P_{Rx}^{eq}=n_{Rx}RT/V_g$ need to be determined. However solving both equations for n_{Rx} gives

$$Vx = \frac{P_{Rx}^{eq} V_g}{RT}$$

or

$$P_{Rx}^{eq} = \frac{RTV}{V_g} x = sx \quad .$$

Using this gives the following format for this heterogeneous reaction.

		$aA_{(aq)}$	+	$bB_{(aq)}$	$\rightleftharpoons$	$cC_{(aq)}$	+	$dD_{(g)}$
lab	in	$[A]_o$		$[B]_o$		$[C]_o$		P_D^o
	eq	$[A]_{eq} = [A]_o - ax$		$[B]_{eq} = [B]_o - bx$		$[C]_{eq} = [C]_o + cx$		$P_D^{eq} = P_D^o + dsx$
		$\frac{1}{a} \downarrow$		$\frac{1}{b} \downarrow$		$\frac{1}{c} \downarrow$		$\frac{1}{d} \downarrow$
LC		$[n_{Rx}^{in}]$		$[n_{Rx}^{in}]$		$[n_{Rx}^{in}]$		P_{Rx}^o

with

$$s = \frac{RTV}{V_g}.$$

The equilibrium quantities would be placed into the equilibrium expression for K_P and the value of x would be determined. Replacing x on the equilibrium line would give all equilibrium quantities. Finally the interpretation on the LC line for reactants is straightforward but both $[n_{Rx}^{in}]$ and P_{Px}^o present on the product side it becomes moe difficult. However one could still determine the number of moles of reaction to find the limiting reagent.

In closing this section the objective has been to treat all equilibrium problems the same way. By asking and answering the three questions, which apply to all problems, you reconstruct the problem into a format that is similar for all problems. The equilibrium expression has the same form for all problems and if it is not obvious whether x is positive or negative the evaluation of the reaction quotient will give you this information. The most difficult part of solving a problem is re-expressing the equilibrium expression in the form of a polynomial of x and using the coefficients in the quadratic formula to find the physical root. The exercises below will help you to become proficient in this somewhat tedious process of solving equilibrium problems. You are encouraged to work as many problems as possible. Doing this will also set the stage for the next section.

EXERCISES

1) For the following reactions, A and B are the reactant molecules and C and D are the product molecules along with the stoichiometric coefficients.

$$A_{(aq)} + B_{(aq)} \rightleftharpoons C_{(aq)} + D_{(aq)}$$

i) For initial concentrations of $[A]_o=1.25M$, $[B]_o=0.725M$, $[C]_o=1.45M$ and $[D]_o=0.645M$, determine K_c if the equilibrium concentration of B ($[B]_{eq}$) is 0.438M. (ans. 3.83)

ii) For initial concentrations of $[A]_o=0.255M$, $[B]_o=0.715M$, $[C]_o=2.85M$ and $[D]_o=0.945M$, determine K_c if the equilibrium concentration of D ($[D]_{eq}$) is 0.538M. (ans. 1.77)

iii) For initial concentrations of $[A]_o=1.25M$, $[B]_o=0.725M$, $[C]_o=1.45M$ and $[D]_o=0.645M$, determine all the equilibrium concentrations with $K_c=0.925$. (ans. x=-.0247)

iv) For initial concentrations of $[A]_o=2.25M$, $[B]_o=0.725M$, $[C]_o=0.845M$ and $[D]_o=1.645M$, determine all the equilibrium concentrations with $K_c=0.925$. (ans. x=0.0227)

v) For initial concentrations of $[A]_o=1.25M$, $[B]_o=1.25M$, $[C]_o=0M$ and $[D]_o=0M$, determine all equilibrium concentrations with $K_c=0.925$. (ans. x=.613)

vi) For initial concentrations of $[A]_o=0$, $[B]_o=0M$, $[C]_o=1.45M$ and $[D]_o=0.245M$, determine all equilibrium concentrations with $K_c=0.925$. (ans. x=-.212)

vii) For initial concentrations of $[A]_o=1.25M$, $[B]_o=1.25M$, $[C]_o 0.645M$ and $[D]_o=0.645M$, determine all equilibrium concentrations with $K_c=0.925$. (ans. x=.284)

2) For the following reactions, A and B are the reactant molecules and C is the product molecule along with the stoichiometric coefficients.

$$A_{(aq)} + B_{(aq)} \rightleftharpoons 2C_{(aq)}$$

i) For initial concentrations of $[A]_o=1.25M$, $[B]_o=0.725M$ and $[C]_o=1.45M$, determine K_c if the equilibrium concentration of B ($[B]_{eq}$) is 0.438M. (ans 9.71)

ii) For initial concentrations of $[A]_o=0.255M$, $[B]_o=0.715M$ and $[C]_o=2.85M$, determine K_c if the equilibrium concentration of C ($[C]_{eq}$) is 0.538M. (ans. 0.110)

iii) For initial concentrations of $[A]_o=1.25M$, $[B]_o=0.725M$ and $[C]_o=1.45M$, determine all the equilibrium concentrations with $K_c=0.925$. (ans. x=-.179)

iv) For initial concentrations of $[A]_o=2.25M$, $[B]_o=0.725M$ and $[C]_o=0.845M$, determine all the equilibrium concentrations with $K_c=0.925$. (ans. x=.122)

v) For initial concentrations of $[A]_o=1.25M$, $[B]_o=1.25M$ and $[C]_o=0M$, determine all equilibrium concentrations with $K_c=0.925$. (ans. x=.406)

vi) For initial concentrations of $[A]_o=0$, $[B]_o=0M$ and $[C]_o=1.45M$, determine all equilibrium concentrations with $K_c=0.925$. (ans. x=-.490)

vii) For initial concentrations of $[A]_o=1.25M$, $[B]_o=1.25M$ and $[C]_o=0.645M$
determine all equilibrium concentrations with $K_c=0.925$. (ans. x=.188)

3) For the following reactions,

$$A_{(g)} + B_{(g)} \rightleftharpoons C_{(g)} + D_{(g)}$$

determine the initial molarities and partial pressures if 0.243mol of A, 0.462mol of B, 0.387mol of C and 0.178mol of D were added to a 2.3L vessel at a temperature of 326K. Determine the equilibrium molarities, partial pressures and the total pressure for $K_P=0.425$. (ans. x=-.290)

Calculator Evaluation of the Equilibrium Expression

In the previous section a general treatment of chemical equilibrium was presented and applied to problems with an equilibrium expression that could be expressed as quadratic polynomial of x. It was seen that it took some effort to get the coefficients in the quadratic polynomial so that x could be found from the quadratic equation. Hopefully you have done many if not all of the previous exercises and feel comfortable with the mathematical manipulations needed to solve equilibrium problems. You might also have come to the conclusion that due to these manipulations this area of Chemistry is not very appealing. I would like you to now consider the following. Suppose you had a friend that when given the answers to the three questions would correctly solve the problem for you (fat chance!). Furthermore this friend would be allowed to accompany you on exams. In essence this friend would make your life a lot easier and this area of Chemistry much more appealing. In actuality you do have a friend like this and this friend is your calculator. Your calculator has some very attractive software that will perform the functions that your friend did. In this section we will introduce this software and show how it will solve these problems for us.

This software is available on many good quality scientific calculators and is called "solver". All TI 80's series (TI-83, TI-84, TI-86 and TI-89) have a solver (for the TI-89 use the numeric solver) and I will only discuss these TI calculators. A solver needs a well-defined mathematical equation that is usually set equal to zero. This equation can have numerous variables. When all but one variable is given a numerical value the solver will tell you what the numerical value of the remaining variable is. If you consider the problems in the previous section, if you told the solver the equilibrium constant and the initial concentrations it would tell you the value of x, which is exactly what your friend would do.

Let us start with the well-defined mathematical expression. The vast majority of equilibrium problems will have reactions that have no more than two reactants or products so that treating the case of two reactants A and B and two products C and D with corresponding stoichiometric coefficients a, b, c and is considered. The general format for this situation is shown again on the next page.

		$aA_{(aq)}$	+	$bB_{(aq)}$	$\rightleftharpoons$	$cC_{(aq)}$	+	$dD_{(aq)}$
lab	in	$[A]_o, V_A$		$[B]_o, V_B$		$[C]_o, V_C$		$[D]_o, V_D$
	eq	$[A]_{eq}=[A]'_o - ax$		$[B]_{eq}=[B]'_o - bx$		$[C]_{eq}=[C]'_o + cx$		$[D]_{eq}=[D]'_o + dx$
		$\frac{1}{a}\downarrow$		$\frac{1}{b}\downarrow$		$\frac{1}{c}\downarrow$		$\frac{1}{d}\downarrow$
LC		$\left[n_{Rx}^{in}\right]$		$\left[n_{Rx}^{in}\right]$		$\left[n_{Rx}^{in}\right]$		$\left[n_{Rx}^{in}\right]$

The general equation for the equilibrium expression given by Eq. (6) is

$$K_c = \frac{\left([C]'_o + cx\right)^c \left([D]'_o + dx\right)^d}{\left([A]'_o - ax\right)^a \left([B]'_o - bx\right)^b}$$

and rearrangement gives

$$K_c \left([A]'_o - ax\right)^a \left([B]'_o - bx\right)^b - \left([C]'_o + cx\right)^c \left([D]'_o + dx\right)^d = 0 \quad .$$

There are ten variables (four initial concentrations, four stoichiometric coefficients, K_c and x) in this equation and this is the mathematical expression to be entered into the calculator. Since this will be used many times you want this to be saved in the calculator. Entering this equation as a y variable can do this. All calculators have a y= key except the TI-86. Pressing this key gives a display of the y variables, which is a list from y1 through yn (n=ten on the TI-83 and TI-84 and n=99 on the TI-89). For the TI-86 press the graph key and select y(x)= and a list of eleven y variables appears. One must also take into consideration how variables can be represented on the calculator in that the variables in the above equation can't be the ones on the calculator. For a TI-83 or TI-84 variables can be represented only with a single letter. For these two calculators the following table gives the relation between the variables.

Symbol in Equation	Symbol on Calculator
K_c	K
$[A]'_o$	A
a	B
$[B]'_o$	C
b	D
$[C]'_o$	E
c	F
$[D]'_o$	G
d	H

The expression entered into the calculator is

$$\backslash Y_1 = (\text{list}) + K*(A-B*X)\wedge B*(C-D*X)\wedge D -$$
$$(E+F*X)\wedge F \; *(G+H*X)\wedge H$$

The solver will display a list of all variables and their order is given by where they appear in the equation (i.e. K,A,B,X.C,D,E,F,G,H). I prefer to use my own order and this is done with (list). The term (list) is just 0*A*B*C*D*E*F*G*H*K*X which does not change the value of the function but the list of variables now becomes A,B,C,D,E,F,G,H,K,X. This makes entering data a little easier. For the TI-86 variables can be combinations of letters and numbers. The table below gives the correspondence between the two sets of variables.

Symbol in Equation	Symbol on Calculator
K_c	K
$[A]'_o$	AI
a	A
$[B]'_o$	BI
b	B
$[C]'_o$	CI
c	C
$[D]'_o$	DI
d	D

With this notation the equation that is entered is (the variable x is gotten by pushing F1)

$$\backslash Y_1 = (\text{list}) + K*(AI - A*x)^\wedge A*(BI - B*x)^\wedge B -$$
$$(CI + C*x)^\wedge C \ *(DI + D*x)^\wedge D$$

where (list) is 0*AI*A*BI*B*CI*C*DI*D*K*x. For the TI-89 only combinations of letters can be used as variables and the table below gives the correspondence between the two sets of variables.

Symbol in Equation	Symbol on Calculator
K_c	k
$[A]'_o$	ai
a	a
$[B]'_o$	bi
b	b
$[C]'_o$	ci
c	c
$[D]'_o$	di
d	d

With this notation the equation that is entered is

$$y1(x) = (\text{list}) + k*(ai - a*x)^\wedge a*(bi - b*x)^\wedge b -$$
$$(ci + c*x)^\wedge c \ *(di + d*x)^\wedge d$$

Where (list) is 0*ai*a*bi*b*ci*c*di*d*k*x. It is very important to understand what the variables on the calculator represent and you should become very comfortable with entering the proper values for the variables.

Now that the well-defined mathematical expression has been permanently entered into the calculator it is now necessary to call the solver. For the TI-83 and TI-84 pushing the math key and scrolling down to the last option, which is the solver, gets the solver. Push enter and the solver appears with a list of the variables and the function that it last used. If the function is not the one wanted scroll up to the function line and push clear. To get the needed y variable push the VARS key, go to

Y-VARS, select Function and press enter. A list of all of the y variables appears and scroll to the one you that want to solve. Push enter and that function will appear in the solver. Push enter again and a list of the variables for that function will appear. For the TI-86 press the solver key and the last function it solved appears at the top of the display. If it is not the function you want all defined functions are shown at the bottom of the display. Press clear and press the more key until the one you want appears. Press the F key underneath it and it will appear. Press enter and a list of the variables will appear. For the TI-89 press the Y= key, go to the desired function and copy it. Not all TI-89's are the same but on mine the numeric solver is gotten by pushing the APPS key, selecting the numeric solver and pushing enter. The last function it used and a list of its variables appears. If it is not the one you want go to the exp line, push clear and paste in the function you copied. Push enter and a list of the variables appears. Note that once a function has been entered into the solver it will remain there until you change it.

Now that the function is on the calculator and the solver has given a list of its variables it is now time to enter data into the list. In some of the examples of the previous section there were two reactants and two products in the equilibrium expression and in other examples there was one reactant and two products in the equilibrium expression. Obviously it would be nice to do both cases with the same expression and this can easily be done. Since any number to the power zero is one, if a reactant or product is missing, set its stoichiometric coefficient equal to zero and use one for its initial concentration (this is just a convenient number to use). Doing this will remove it from the equilibrium expression. For all examples where B was a solvent b=0 and $[B]'_o=1$ which gives the equilibrium expression preceding Eq. (9).

Suppose now that all initial concentrations and K_c have been properly entered into the list and you would like to determine x. For all calculators move the cursor to x. The value of x is determined by using a numerical algorithm and the better the starting guess that you enter the easier it will find the correct value of x. Since you have already found whether x is positive or negative you know whether to use a − guess or a +guess. A reasonable value of guess can be obtained from the LC line. If the reaction goes to the product side x must be between 0 and the smallest entry on the LC line on the reactant side. If the reaction goes to the reactant side x must be between zero and minus the smallest entry on the LC line on the product side. Halfway between the ranges often works well but consideration of K_c is also helpful to choose a good guess for x (if K_c is very small and the reaction goes to the product side a small value for the guess would work better than half the LC line value). Keep in mind that polynomials often have both positive and negative roots so that this ensures that you will obtain the correct physical root. Enter in the guess for x. For the TI-83 or TI-84 press the solve option located on the enter key. For the TI-86 push the F key under the solve option on the menu list at the bottom of the display. For the TI-89 press the F key for the solve option in the menu list at the top of the display.

Before doing some examples it is worthwhile to point out another at times helpful feature, which is the bound line. The bound line limits where the solver will look for the value of x. The default values correspond to minus a large number to plus a large number. When the solver is opened it will always have these default values. In cases when x is a very small positive or negative number it is helpful to change the values in bound to help the solver find x. If x is a positive very small number it is helpful to change the lower bound to a positive number that is even much smaller (something like 10^{-40}) and use a very small guess for x (something like 10^{-7}). If x is a very small negative number change the upper bound to something very small (-10^{-40}) and a very small negative guess (-10^{-7}). In short the better the guess is and using a helpful bound can really help in finding x.

We will now re-do the example problems of the previous section using the solver. This will help to develop the techniques needed to completely solve a problem without ever writing an equation on a piece of paper. It is still very helpful to use the format so that all of the measurements and objectives in the problem are well defined. In the first example the measurements were $[A]_o=1.0M$, $[B]_o=1.5M$, $[C]_o=1.3M$, $[D]_o=0.25M$ and $[D]_{eq}=0.65M$ and the objective was to determine the equilibrium constant. This gave the format below.

		$A_{(aq)}$	+	$B_{(aq)}$	$\rightleftharpoons$	$C_{(aq)}$	+	$D_{(aq)}$
lab	in	1.0		1.5		1.3		.25
	eq	$[A]_{eq} = 1-x$		$[B]_{eq} = 1.5-x$		$[C]_{eq} = 1.3+x$		$.65 = .25+x$

LC

The procedure I use for solving this type of problem with the solver has the following steps.

i) Enter the initial concentrations and stoichiometric coefficients into the list.
ii) With the help of the format go to the X line and determine X
(here just type in .65-.25 on the X line, move the cursor and .4 appears)
and now X is a well-defined variable.
iii) Use X to change the initial concentrations into equilibrium concentrations. Subtract the stoichiometric coefficient times X for the reactants and add the stoichiometric coefficient times X for the products. Note than using the variable X rather than the actual number for X makes this easier.
iv) Go to the X line and set it equal to zero.
v) Go to the K line and solve for K.

For these problems I will illustrate the solutions for the TI-83 calculator since it seems to be most popular. The only changes for the other calculators are for what variables are called. In the notation below a number arrow letter meant that the number was entered into the list for the letter. The steps above are now shown.

i) The initial values and stoichiometric coefficients are entered into the list.
$1 \to A\ 1 \to B\ 1.5 \to C\ 1 \to D\ 1.3 \to E\ 1 \to F\ .25 \to G\ 1 \to H\ \ \to K\ \ \to X$
The list becomes as shown below.
A = 1
B = 1
C = 1.5
D = 1
E = 1.3
F = 1
G = .25
H = 1
K = 0
X = 0

ii) On the X line type in .65-.25 and move cursor. This gives the list below.
A = 1
B = 1
C = 1.5
D = 1
E = 1.3
F = 1
G = .25
H = 1
K = 0
X = .4

iii) Subtract X from A (1-X) and C (1.5-X) and add X to E (1.3+X) and G (.25+X) giving the list.
A = .6
B = 1
C = 1.1
D = 1
E = 1.7
F = 1
G = .65
H = 1
K = 0
X = .4

iv) Setting X to zero gives the list.   v) Solving for K gives the list.

        A = .6   A = .6
        B = 1   B = 1
        C = 1.1   C = 1.1
        D = 1   D = 1
        E = 1.7   E = 1.7
        F = 1   F = 1
        G = .65   G = .65
        H = 1   H = 1
        K = 0   K = 1.6742....
        X = 0   X = 0

Rounding off gives the answer of K_c=1.7. Note that no equations had to be written.

    The second example had the measurements of K_c=1.7, $[A]_o$=.25M, $[B]_o$=.57M, $[C]_o$=1.2M and $[D]_o$=.85M and the objectives of determining all equilibrium concentrations. The format with the measurements and objectives is given below.

$$A_{(aq)} + B_{(aq)} \rightleftharpoons C_{(aq)} + D_{(aq)}$$

lab in   .25   .57   1.2   .85
    eq $[A]_{eq} = .25 - x$ $[B]_{eq} = .57 - x$ $[C]_{eq} = 1.2 + x$ $[D]_{eq} = .85 + x$

LC   .25   .57   1.2   .85

The procedure I use for solving this type of problem with the solver has the following steps.
i)   Enter the initial concentrations, the stoichiometric coefficients and K_c into the list.
ii) If necessary (as in this case) determine the reaction quotient. This is easily done by setting X equal to zero in the list and solving for K (for x=0 the equilibrium expression becomes the reaction quotient). If K is greater than K_c, X is negative and if K is less than K_c, X is positive.
iii) Enter K_c back into the list, go to X, enter a good guess and solve for X.

iv) Use X to determine the equilibrium concentrations. Add the appropriate stoichiometric coefficient times X to the products and subtract the appropriate stoichiometric coefficient times X to the reactants.

v) Set X equal to zero and solve for K which if the problem was done correctly should be K_c (a five second check).

The steps above are now shown.

i) The initial values, stoichiometric coefficients and K_c are entered as shown.
.25 → A 1 → B .57 → C 1 → D 1.2 → E 1 → F .85 → G 1 → H 1.7 → K   → X

The list becomes as shown below.   ii)   Set X=0, solve for K and list becomes as shown below.

A = .25	A = .25
B = 1	B = 1
C = .57	C = .57
D = 1	D = 1
E = 1.2	E = 1.2
F = 1	F = 1
G = .85	G = .85
H = 1	H = 1
K = 1.7	K = 7.15789...
X = 0	X = 0

iii) Set K to 1.7 and, since x<0, .85 on LC line is smallest so try X=−.4 and solve for X giving the list below.

A = .25
B = 1
C = .57
D = 1
E = 1.2
F = 1
G = .85
H = 1
K = 1.7
X = −.216316...

iv) Determining equilibrium value gives list below.

A = .4663167....
B = 1
C = .7863167....
D = 1
E = .9836832....
F = 1
G = .6336832...
H = 1
K = 1.7
X = −.216316...

v) Set X=0 and solve for K to see if solution was correct which gives the list below.

A = .4663167....
B = 1
C = .7863167....
D = 1
E = .9836832....
F = 1
G = .6336832...
H = 1
K = 1.7
X = 0

Of course the answers reported should be expressed with the proper number of significant figures (two in this example). One should note that no equations were written down in order to solve this problem which is hardly the case for the previous treatment.

Chapter 4: Chemical Equilibrium | 145

The next example had the measurements of $K_c=1.4\times10^{-7}$ and only A was present initially with $[A]_o=.75$M and the objectives were to determine the equilibrium concentrations. This gave the following format.

		$A_{(aq)}$	+	$B_{(\ell)}$	$\rightleftharpoons$	$C_{(aq)}$	+	$D_{(aq)}$
lab	in	.75		xs		0		0
	eq	$[A]_{eq}=.75-x$				$[C]_{eq}=x$		$[D]_{eq}=x$
	LC	.75				0		0

The steps are now shown below. (Note that ii) is not needed since x must be positive)

i) The initial values and K_c are entered as shown. Since B is not in the equilibrium expression its initial concentration (C) is one and its stoichiometric coefficient (D) is zero.

$.75 \rightarrow A\ 1 \rightarrow B\ 1 \rightarrow C\ 0 \rightarrow D\ 0 \rightarrow E\ 1 \rightarrow F\ 0 \rightarrow G\ 1 \rightarrow H\ 1.4E-7 \rightarrow K \quad \rightarrow X$

The list becomes as shown below.

A = .75

B = 1

C = 1

D = 0

E = 0

F = 1

G = 0

H = 1

K = 1.4E – 7

X = 0

146 | Stoichiometry and Beyond

iii) K_c is very small so that X should be small so try X=.01 and solve for X which gives list below.

    A = .75

    B = 1

    C = 1

    D = 0

    E = 0

    F = 1

    G = 0

    H = 1

    K = 1.4E – 7

    X = 3.23967...E – 4

iv) Determining equilibrium value gives list below.

    A = .74967603...

    B = 1

    C = 1

    D = 0

    E = 3.23967...E – 4

    F = 1

    G = 3.23967...E – 4

    H = 1

    K = 1.4E – 7

    X = 3.23967...E – 4

v) Set X=0 and solve for K to see if solution was correct which gives the list below.

    A = .74967603...

    B = 1

    C = 1

    D = 0

    E = 3.23967...E – 4

    F = 1

    G = 3.23967...E – 4

    H = 1

    K = 1.3999999892...E – 7

    X = 3.23967...E – 4

Of course this problem could have been solved using the simple approximate treatment previously described but the calculator solution is still simple, doesn't require any writing of equations and gives the same results. In my opinion I would not like to worry whether approximations will work. I would rather just solve the problem.

The next example had the measurements $K_c=1.4 \times 10^{-7}$, $[A]_o=0M$, $[C]_o=.75M$ and $[D]_o=.75M$ and the objectives were to determine the equilibrium concentrations. Using this gives the following format.

		$A_{(aq)}$	+	$B_{(\ell)}$	⇌	$C_{(aq)}$	+	$D_{(aq)}$
lab	in	0		xs		.75		.75
	eq	$[A]_{eq} = -x$				$[C]_{eq} = .75 + x$		$[D]_{eq} = .75 + x$
	LC	0				.75		.75

The steps are now shown below. (Note that ii) is not needed since x must be negative)

i) The initial values and K_c are entered as shown.
$0 \to A\ 1 \to B\ 1 \to C\ 0 \to D\ .75 \to E\ 1 \to F\ .75 \to G\ 1 \to H\ 1.4E-7 \to K\ \to X$
The list becomes as shown below.
A = 0
B = 1
C = 1
D = 0
E = .75
F = 1
G = .75
H = 1
K = 1.4E − 7
X = 0

iii) The value of .75 is on the LC line so try X=−.4 and solve for X which gives list below.
A = 0
B = 1
C = 1
D = 0
E = .75
F = 1
G = .75
H = 1
K = 1.4E − 7
X = −.749676...

iv) Determining equilibrium value gives list below.
A = .74967603...
B = 1
C = 1
D = 0
E = 3.23967...E − 4
F = 1
G = 3.23967...E − 4
H = 1
K = 1.4E − 7
X = −.749676...

v) Set X=0 and solve for K to see if solution was correct which gives the list below.

$A = .74967603...$

$B = 1$

$C = 1$

$D = 0$

$E = 3.23967...E - 4$

$F = 1$

$G = 3.23967...E - 4$

$H = 1$

$K = 1.3999999892...E - 7$

$X = 3.23967...E - 4$

The final example had measurements of $K_c = 2.7 \times 10^{-10}$ and $[A]_o = 0M$, $[C]_o = .74M$ and $[D]_o = 1.3M$ and the objectives were to determine the equilibrium concentrations. Using this gives the format below.

		$A_{(aq)}$	+	$B_{(\ell)}$	⇌	$C_{(aq)}$	+	$D_{(aq)}$
lab	in	0		xs		.74		1.3
	eq	$[A]_{eq} = -x$				$[C]_{eq} = .74 + x$		$[D]_{eq} = 1.3 + x$
	LC					.74		1.3

The steps are now shown below. (Note that ii) is not needed since x must be negative)

i) The initial values and K_c are entered as shown.
$0 \to A\ 1 \to B\ 1 \to C\ 0 \to D\ .74 \to E\ 1 \to F\ 1.3 \to G\ 1 \to H\ 2.7E-10 \to K\ \to X$
The list becomes as shown below.
A = 0
B = 1
C = 1
D = 0
E = .74
F = 1
G = 1.3
H = 1
K = 2.7E – 10
X = 0

iii) The value of .74 is on the LC line so try X=-.4 and solve for X which gives list below.
A = 0
B = 1
C = 1
D = 0
E = .74
F = 1
G = 1.3
H = 1
K = 2.7E – 10
X = –.73999999964321

iv) Determining equilibrium value gives list below.
A = .7399999...
B = 1
C = 1
D = 0
E = 3.5679E – 10
F = 1
G = .5600000...
H = 1
K = 2.7E – 10
X = –.73999999964321

v)   Set X=0 and solve for K to see if solution was correct which gives the list below.

   A = .7399999...
   B = 1
   C = 1
   D = 0
   E = 3.5679E – 10
   F = 1
   G = .5600000...
   H = 1
   K = 2.7000324...E – 10
   X = 0

In this example the actual solver number for X was given and you can see that this number has a precision of fourteen digits to the right of the decimal point. This gives plenty of accuracy to solve most problems without making any modification what initial state the reaction begins in as was done before.

EXERCISES

Do the evaluation for the problems in the previous section using the solver techniques developed here.

Some more examples are now done to further demonstrate the capabilities of the solver. Consider the format for the following reaction with the defined initial concentrations and stoichiometric coefficients given below and K_c=0.0125. This is a fifth order polynomial of x but this is not a problem for the solver.

		$A_{(aq)}$ +	$3B_{(\ell)}$	⇌	$2C_{(aq)}$ +	$3D_{(aq)}$
lab	in	.515	.715		.527	.182
	eq	$[A]_{eq} = .515 - x$	$[B]_{eq} = .715 - 3x$		$[C]_{eq} = .527 + 2x$	$[D]_{eq} = .182 + 3x$
	LC	.515	.238		.264	.607

The steps for solving this problem are shown below.

i) The initial values, stoichiometric coefficients and K_c are entered as shown.
.515 → A 1 → B .715 → C 3 → D .537 → E 2 → F .182 → G 3 → H .0125 → K   → X

The list becomes as shown below.

ii) Set X=0, solve for K and list becomes as shown below.

A = .515
B = 1
C = .715
D = 3
E = .537
F = 2
G = .182
H = 3
K = .0125
X = 0

A = .515
B = 1
C = .715
D = 3
E = .537
F = 2
G = .182
H = 3
K = .009235...
X = 0

iii) Set K to .0125 and, since x>0, .515 on LC line is smallest so try X=.2 and solve for X giving the list below.

A = .515
B = 1
C = .715
D = 3
E = .537
F = 2
G = .182
H = 3
K = .0125
X = .0043367...

iv) Determining equilibrium value gives list below.

A = .5106632...
B = 1
C = .7019898...
D = 3
E = .54567...
F = 2
G = .19501...
H = 3
K = .0125
X = .0043367...

152 | Stoichiometry and Beyond

v) Set X=0 and solve for K to see if solution was correct which gives the list below.

A = .5106632...

B = 1

C = .7019898...

D = 3

E = .54567...

F = 2

G = .19501...

H = 3

K = .012499999...

X = 0

As one can see the solution of these more complicated reactions is no more difficult than the class of reactions that can be solved with the quadratic formula.

Before closing this section is helpful to consider what modifications are necessary for reactions with more than two reactants or products. With another compound in a reaction there would be another term in the equilibrium expression. This compound could be represented with the symbols I and J for the TI-83 or TI-84, the symbols EI and E for the TI-86 or the symbols ei and e for the TI-89. If the compound is a reactant there would be (I-J*X)^J in the term multiplying K on the TI-83 or if it were a product it would be in the other term. This is easily accomplished by editing the expression using the INS key. The other calculators are modified in an analogous way. As indicated in the previous section another modification is needed for heterogeneous reactions. As an example of this consider the format for the reaction below at 298K, K_P=14.8, V_ℓ=0.100L, V_g=0.900L with the initial amounts shown below in the format below. The LC line is omitted because its interpretation for heterogeneous systems is not as straightforward.

	$A_{(aq)}$	+	$3B_{(g)}$	⇌	$2C_{(g)}$	+	$3D_{(aq)}$
init.	1.43M		0.719atm		0.973atm		1.56M
equil.	1.43 − x		0.719 − 3sx		0.973 + 2sx		1.56 + 3x

Since s=RTV_ℓ/V_g, this can be included in the equilibrium expression for both the second reactant and the second product by editing the expression and using the INS key. The expression would become K*(A-B*X)^B*(C-D*S*X)^D-(E-F*S*X)^F*(G-H*X)^H for the TI-83 and there would be analogous expressions for the other calculators.

The steps for solving this problem are shown below.

i) The initial values, stoichiometric coefficients, S and K_c are entered as shown.
1.43 → A 1 → B .719 → C 3 → D .973 → E 2 → F 1.56 → G 3 → H 14.8 → K   → X

The list becomes as shown below.

A = 1.43
B = 1
C = .719
D = 3
E = .937
F = 2
G = 1.56
H = 3
K = 14.8
X = 0
S = .082057 * 298 * .1 / .9

ii) Set X=0, solve for K and list becomes as shown below.

A = 1.43
B = 1
C = .719
D = 3
E = .937
F = 2
G = 1.56
H = 3
K = 6.2709...
X = 0
S = 2.71699...

iii) Set K to 14.8 and, since x>0, try X=.5 and solve for X giving the list below.

A = 1.43
B = 1
C = .719
D = 3
E = .937
F = 2
G = 1.56
H = 3
K = 14.8
X = .015634...
S = 2.71699...

iv) Determining equilibrium values gives list below.

A = 1.41436...
B = 1
C = .59156...
D = 3
E = 1.0219...
F = 2
G = 1.6069...
H = 3
K = 14.8
X = .015634...
S = 2.71699...

v)   Set X=0 and solve for K to see if solution was correct which gives the list below.

   A = 1.41436...
   B = 1
   C = .59156...
   D = 3
   E = 1.0219...
   F = 2
   G = 1.6069...
   H = 3
   K = 14.799999...
   X = 0
   S = 2.71699...

Even heterogeneous reactions can be treated so that equilibrium quantities for virtually any reaction can be obtained without much effort by using the solver. Hopefully you now see that you do have a real friend in this area of Chemistry.

EXERCISES

1)   For the following reactions, A and B are the reactant molecules and C and D are the product molecules along with the stoichiometric coefficients.

$$2A_{(aq)} + 3B_{(aq)} \rightleftharpoons 4C_{(aq)} + 2D_{(aq)}$$

   i)   For initial concentrations of $[A]_o=1.25M$, $[B]_o=0.725M$, $[C]_o=1.45M$ and $[D]_o=0.645M$, determine K_c if the equilibrium concentration of B ($[B]_{eq}$) is 0.438M. (ans. 83.8)

   ii) For initial concentrations of $[A]_o=0.255M$, $[B]_o=0.715M$, $[C]_o=2.85M$ and $[D]_o=0.945M$, determine K_c if the equilibrium concentration of D ($[D]_{eq}$) is 0.538M. (ans. 4.87)

iii) For initial concentrations of $[A]_o=1.25M$, $[B]_o=0.725M$, $[C]_o=1.45M$ and $[D]_o=0.645M$, determine all the equilibrium concentrations with $K_c=0.925$. (ans. x=-.114)

iv) For initial concentrations of $[A]_o=2.25M$, $[B]_o=0.725M$, $[C]_o=0.845M$ and $[D]_o=1.65M$, determine all the equilibrium concentrations with $K_c=0.925$. (ans. x=0.0603)

v) For initial concentrations of $[A]_o=1.25M$, $[B]_o=1.25M$, $[C]_o=0M$ and $[D]_o=0M$, determine all equilibrium concentrations with $K_c=0.925$. (ans. x=.625)

vi) For initial concentrations of $[A]_o=0$, $[B]_o=0M$, $[C]_o=1.45M$ and $[D]_o=0.245M$, determine all equilibrium concentrations with $K_c=0.925$. (ans. x=-.107)

vii) For initial concentrations of $[A]_o=1.25M$, $[B]_o=1.25M$, $[C]_o 0.645M$ and $[D]_o=0.645M$, determine all equilibrium concentrations with $K_c=0.925$. (ans. x=.113)

2) For the following reactions, A and B are the reactant molecules and C is the product molecule along with the stoichiometric coefficients.

$$2A_{(aq)} + B_{(aq)} \rightleftharpoons 3C_{(aq)}$$

i) For initial concentrations of $[A]_o=1.25M$, $[B]_o=0.725M$ and $[C]_o=1.45M$, determine K_c if the equilibrium concentration of B ($[B]_{eq}$) is 0.438M. (ans 61.7)

ii) For initial concentrations of $[A]_o=0.255M$, $[B]_o=0.715M$ and $[C]_o=2.85M$, determine K_c if the equilibrium concentration of C ($[C]_{eq}$) is 0.538M. (ans. 0.0325)

iii) For initial concentrations of $[A]_o=1.25M$, $[B]_o=0.725M$ and $[C]_o=1.45M$, determine all the equilibrium concentrations with $K_c=0.925$. (ans. x=-.0955)

iv) For initial concentrations of $[A]_o=2.25M$, $[B]_o=0.725M$ and $[C]_o=0.845M$, determine all the equilibrium concentrations with $K_c=0.925$. (ans. x=.145)

v) For initial concentrations of $[A]_o=1.25M$, $[B]_o=1.25M$ and $[C]_o=0M$, determine all equilibrium concentrations with $K_c=0.925$. (ans. x=.263)

vi) For initial concentrations of $[A]_o=0$, $[B]_o=0M$ and $[C]_o=1.45M$, determine all equilibrium concentrations with $K_c=0.925$. (ans. x=-.318)

vii) For initial concentrations of $[A]_o=1.25M$, $[B]_o=1.25M$ and $[C]_o=0.645M$ determine all equilibrium concentrations with $K_c=0.925$. (ans. x=.125)

3) For the following reactions,

$$4A_{(g)} + 3B_{(g)} \rightleftharpoons 2C_{(g)} + 3D_{(g)}$$

determine the initial molarities and partial pressures if 0.243mol of A, 0.462mol of B, 0.387mol of C and 0.178mol of D were added to a 2.3L vessel at a temperature of 326K. Determine the equilibrium molarities, partial pressures and the total pressure for $K_P=0.425$. (ans. x=.237)

Applications of Le Châtelier's Principle

A very valuable tool called Le Châtelier's Principle for making qualitative predictions to how a system in dynamic equilibrium will respond to a disturbance is now introduced. This principle says the following.

When a system that is in a state of dynamic equilibrium is disturbed, the system will respond in a way that counteracts the disturbance.

This section will use this principle to predict how a reaction in a state of dynamic equilibrium responds to various disturbances. When a reaction is in a state of dynamic equilibrium, the rate of the forward reaction is the same as the rate of the reverse reaction so that no changes in concentrations or partial pressures of the reactants or products occur. This state of the reaction is written as

$$\text{Reactants} \rightleftharpoons \text{Products} \quad .$$

There are a variety of disturbances that are now considered.

Consider the response when either a reactant or a product that is in the equilibrium expression is added. According to Le Châtelier's principle the response is to remove the reactant or product that was added. This causes a shift to the product side if a reactant is added or to the reactant side if a product is added. This is summarized below

$$\text{Reactants} \xrightarrow{\text{add reactant}} \text{Products}$$

$$\text{Reactants} \xleftarrow{\text{add product}} \text{Products}$$

where the arrow indicates the shift and the disturbance is above the arrow. Note that adding a reactant that is a pure solid or in most cases a pure liquid does not cause a shift since it is not in the equilibrium expression.

Consider now the response when heat is either added or removed. According to Le Châtelier's Principle, the response is to remove heat if heat is added or to add heat if heat is removed. By convention, the ΔH_{Rx} of a reaction with units of kJ/mol is defined as the enthalpy change when one mole of reaction from the reactant side to the product side occurs. For the forward reaction one has

Chapter 4: Chemical Equilibrium | 159

$$\text{Reactants} \xrightarrow{\Delta H_{Rx}} \text{Products}$$

and, since the enthalpy change changes its sign for the reverse reaction, for the reverse reaction

$$\text{Products} \xleftarrow{-\Delta H_{Rx}} \text{Reactants} .$$

If $\Delta H_{Rx}<0$, the reaction is called an exothermic reaction and heat is produced by the forward reaction. Is this case $-\Delta H_{Rx}>0$ (which is called an endothermic reaction) so that the reverse reaction consumes heat. If $\Delta H_{Rx}>0$, the forward reaction consumes heat and the reverse reaction produces heat. In other words, knowing the sign of ΔH_{Rx} determines which reaction is exothermic and which reaction is endothermic. For the case that $\Delta H_{Rx}<0$, if heat is added the response is

$$\text{Reactants} \xleftarrow{\text{heat added}} \text{Products}$$

and if heat is removed

$$\text{Reactants} \xrightarrow{\text{heat removed}} \text{Products} .$$

For the case that $\Delta H_{Rx}>0$, if heat is added the response is

$$\text{Reactants} \xrightarrow{\text{heat added}} \text{Products}$$

and if heat is removed

$$\text{Reactants} \xleftarrow{\text{heat removed}} \text{Products} .$$

Another disturbance could be to increase or decrease the volume at constant temperature, which could affect the equilibrium quantities of gases or solutes. Consider first the gas phase reaction below.

$$aA_{(g)} + bB_{(g)} \rightleftarrows cC_{(g)} + dD_{(g)}$$

If the volume of a container having a gas in a state of dynamic equilibrium is increased, the total pressure of the gas would decrease. Therefore the response would be to increase the total pressure. If a+b>c+d, the total pressure would be increased by a shift to the reactant side (the side with more moles of gas). If a+b<c+d, the total pressure would increase by a shift to the product side. If a+b=c+d, no shift would occur. For a decrease in volume (increase in total pressure) the

reaction would shift to the side with lowest sum of stoichiometric coefficients. Notice also that the total pressure of a gas in a state of dynamic equilibrium could be increased by adding an inert gas (a gas that does not react). However this would not change the equilibrium values of the partial pressures so that no shift would occur.

Consider now a reaction in the liquid phase consisting of solute molecules in a solvent that is assumed to be water.

$$aA_{(aq)} + bB_{(aq)} \rightleftharpoons cC_{(aq)} + dD_{(aq)}$$

The equilibrium expression contains only the equilibrium molarities of the solutes, which are the equilibrium number of moles of the solutes divided by the total volume. Suppose now that some solvent is added to the solution, which would increase the total volume. One could use the equilibrium number of moles divided by the new volume of each solute to determine the reaction quotient which would, when compared to the equilibrium constant, indicate which way the reaction would shift. A more direct way would be to add together all of the equilibrium molarities to obtain the total molarity of all solutes. When the solvent is added the total molarity would decrease. Therefore the response would be to increase the total molarity. If $a+b>c+d$, the total molarity would increase by a shift to the reactant side (the side that has more moles of solutes). If $a+b<c+d$, the shift would be to the product side. If $a+b=c+d$, there would be no shift. If some solvent is removed (increase in total molarity) the shift (if any) would be to the side with the fewest moles of solutes.

Finally a catalyst is not used up in a chemical reaction so that it is not present in the equilibrium expression. Therefore adding a catalyst does not affect the equilibrium quantities. The purpose of a catalyst is to increase the rate of the reaction not the final amounts of the reactants and products.

EXERCISES

1) For the following reaction $\Delta H_{Rx}<0$. For each disturbance, indicate which way (if any) the reaction will shift.

$$A_{(g)} + 3B_{(g)} \rightleftharpoons 2C_{(g)} + 3D_{(g)}$$

i) A is added.

ii) B is removed.

iii)   C is removed.

iv)   D is added.

v)   Heat is removed.

vi)   Heat is added.

vii)   Volume is increased.

viii) Volume is decreased.

ix)   An inert gas is added.

2) For the following reaction $\Delta H_{Rx}>0$. For each disturbance, indicate which way (if any) the reaction will shift.

$$A_{(aq)} + 2B_{(aq)} \rightleftharpoons C_{(aq)} + D_{(aq)}$$

i)   A is added.

ii)   B is removed.

iii)   C is removed.

iv)   D is added.

v)   Heat is removed.

vi)   Heat is added.

vii)   Solvent is added.

viii) Solvent is removed.

Index

absolute temperature, 94
acid-base reactions, 77
activity, 107
actual yield, 60
Alchemy, 1
alpha rays, 17
Arrhenius definition, 84
atomic mass, 5, 20, 21, 23, 24, 25, 26, 27, 29
atomic units, 21
average atomic mass, 21
Avogadro's number, 31, 36
Avogadro's Law, 95
balanced reaction, 13, 48, 51, 77, 100
beta ray, 17, 20
Boyle's Law, 94
Bronsted and Lowry Theory, 84, 85
Calculator Evaluation, 136
calculators
    types of, 136
    variables for, 137
calling the solver, 139
charge to mass ratio, 18
Charles's Law, 94
chemical equilibrium, 51, 105, 136
combustion, 2, 42
compound, *definition*, 3
conversion factors, 13, 24, 32, 49, 55, 79
Crookes tube, 18
Dalton's Law of Partial Pressures, 98
Dalton's theory, 9, 11, 22
Dalton's Theory, 2, 7, 17
Dalton's Theory (revised), 26
Dalton's theory, 4
dilution, 71, 119
diprotic acid, 86
dynamic equilibrium, 83, 84, 105, 159
electron, 18, 41
empirical formula, 32
Empirical formula, 37
end point, 89
entering data into solver, 138
equilibrium constant, K_c, 106
equilibrium constant, K_P, 108
equilibrium expression, 115, 121, 128, 136, 140, 153, 159
Equilibrium Format, 115
equilibrium problems, 131
    solving, 121
    solving with a calculator, 136
equivalence point, 88
format, 28, 32, 37, 43, 47, 51
gamma ray, 17
Gay-Lussac's interpretation of the Law of Combining Volumes, 8
group displacement laws, 18, 20
half-life, 17
heterogeneous reaction, 115, 130, 153, 155
heterogeneous reactions
    solving equilibrium problems for, 131
homogeneous reaction, 115
ideal gas equation of state, 97
isotope, 20, 23, 27
isotopes
    notation, 20
lab and mole lines, 115
    moving between, 52, 61, 72, 81
Law of Combining Volumes, 8, 99
Law of Conservation of Mass, 2, 5
Law of Constant Composition, 2, 5, 16
Law of Multiple Proportions, 2, 6
limiting reagent, 47, 50, 52, 57, 62, 74, 80, 101, 119, 128, 131
logical relations, 13, 36, 49, 55, 58
matter
    theories of, 1
molar mass, 14, 21, 32, 34, 38, 43, 54, 61
molarity, 71, 88, 107, 161
Mole Concept, 30
mole world, 98
    moving in, 52, 55, 71, 75
monoprotic acid, 85, 89
net ionic reaction, 78, 82, 87
neutralization reaction, 88
neutron, 20, 27
nucleus, 19, 22
oil drop experiment, 18
percent abundance, 23
percent mass, 3, 16, 36, 39, 70
percent yield, 60, 76, 105, 127
polynomials of higher order
    solving equilibrium problems with, 105, 120
precipitation reaction, 77, 80
property of matter, *definition*, 95

Index | 163

proton, 20, 84, 86, 89
quadratic formula, 120, 131, 153
quadratic polynomial, 120, 136
radioactivity, 17
reaction quotient, 105, 113, 118, 130, 161
relative atomic mass, 8, 11
rule of greatest simplicity, 8
Rutherford's atomic model, 19
Scientific Method, 1
solubility rules, 79
solver, 136
spectator ion, 78, 82
stoichiometry, 2, 21, 51, 60, 64, 97, 106, 115
Stoichiometry, 47
   for gases, 94
   for masses, 53
   for solutes, 70
theoretical yield, 60
Thomson's atomic model, 19
Three Questions, The, 51, 75, 80, 115, 136
titration, 88
y variable, 137

Printed by Libri Plureos GmbH in Hamburg, Germany